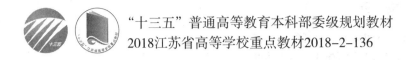

"十三五"普通高等教育本科部委级规划教材

2018江苏省高等学校重点教材2018-2-136

纺织新材料（双语）
New Textile Materials

王春霞　陈嘉毅　主　编
贾高鹏　吴佳佳　副主编

中国纺织出版社有限公司

国家一级出版社
全国百佳图书出版单位

内 容 提 要

本书采用英汉对照的形式，介绍新型纺织纤维，包括纤维素纤维（如木棉纤维、彩棉纤维和竹纤维）、蛋白质纤维（如牛奶蛋白纤维、大豆蛋白纤维和改性羊毛）、合成纤维（如聚酯纤维）及高性能纤维（如碳纤维、玻璃纤维、芳纶、高强聚乙烯纤维）等。通过学习，既可掌握各种新型纤维的发展、制备、结构、性能、应用及存在问题，同时又可提高纺织专业英语水平，为今后进一步学习纺织专业课双语课程打下基础。

本书可作为高等学校纺织工程及相关专业双语教学的入门教材，也可供工程技术人员参考。

图书在版编目（CIP）数据

纺织新材料 = New Textile Materials：汉英对照 / 王春霞，陈嘉毅主编. --北京：中国纺织出版社有限公司，2019.11

"十三五"普通高等教育本科部委级规划教材 2018江苏省高等学校重点教材

ISBN 978-7-5180-6751-0

Ⅰ.①纺… Ⅱ.①王… ②陈… Ⅲ.①纺织纤维—材料科学—双语教学—高等学校—教材—汉、英 Ⅳ.①TS102

中国版本图书馆CIP数据核字（2019）第216298号

责任编辑：符 芬　责任校对：王花妮　责任印制：何 建

中国纺织出版社有限公司出版发行
地址：北京市朝阳区百子湾东里A407号楼　邮政编码：100124
销售电话：010—67004422　传真：010—87155801
http://www.c-textilep.com
E-mail: faxing@c-textilep.com
中国纺织出版社天猫旗舰店
官方微博 http://weibo.com/2119887771
三河市宏盛印务有限公司印刷　各地新华书店经销
2019年10月第1版第1次印刷
开本：787×1092　1/16　印张：13.5
字数：252千字　定价：58.00元

凡购本书，如有缺页、倒页、脱页，由本社图书营销中心调换

Foreword 前言

高校专业课实施双语教学既是高等教育的发展趋势，也是高校教育改革的重要方向。随着专业课双语教学的推广，缺乏适宜的专业教材是主要问题之一。目前，专业课双语教材的使用主要是引进或改编外国原版教材和自编专业双语教材，由于外国版权出口限制，很难引入最新的、系统的、前沿的专业教材，再加上外国原版教材价格昂贵，因此，自编专业双语教材是双语教材本土化的最佳方案。

本书针对高校纺织专业课程双语教学需要编写，目的是掌握专业知识的同时，提高专业英语水平，因此，将纺织新材料作为本书的主体内容，采用中英文对照形式介绍11种新型纤维的发展、制备、结构、性能及产品开发等方面。

本书由盐城工学院纺织服装学院王春霞负责组织团队编写，共同策划和确立本书的大纲和框架，最后由王春霞和陈嘉毅统稿。具体章节的完成情况如下：中文部分第一章至第七章由王春霞完成，第八章至第十一章由贾高鹏完成；英文部分由王春霞、陈嘉毅、贾高鹏和吴佳佳共同完成。本书得到了盐城工学院教材出版基金的资助。

除了上述执笔人员外，季萍、刘丽、陆振乾、高大伟、马志鹏和刘国亮等提供了有关资料和数据，吕立斌、宋孝浜和毕红军等对本书提出了很多宝贵意见，在此一并致谢。另外，书中参考了其他教材和专业资料，在此对参考文献作者谨表示真心感谢。

由于时间仓促，编者水平有限，书中难免有不妥或疏漏之处，恳请广大读者和专家提出宝贵意见。

编者
2019年8月

Contents 目 录

Part One　Cellulose Fiber/ 纤维素纤维

Chapter One　Kapok Fiber/ 木棉纤维 ·········· 2
  1　Introduction/ 前言 ·········· 2
  2　Structure/ 结构 ·········· 3
    2.1　Molecular Structure/ 分子结构 ·········· 3
    2.2　Supramolecular Structure/ 超分子结构 ·········· 4
    2.3　Morphological Structure/ 形态结构 ·········· 4
  3　Property/ 性能 ·········· 6
    3.1　Physical Property/ 物理性能 ·········· 6
    3.2　Chemical Property/ 化学性能 ·········· 9
  4　Application/ 应用 ·········· 11
  Exercises/ 练习 ·········· 13
  References/ 参考文献 ·········· 13

Chapter Two　Natural Colored Cotton Fiber/ 天然彩棉纤维 ·········· 14
  1　Introduction/ 前言 ·········· 14
  2　Structure/ 结构 ·········· 16
    2.1　Molecular Structure/ 分子结构 ·········· 16
    2.2　Supramolecular Structure/ 超分子结构 ·········· 18
    2.3　Morphological Structure/ 形态结构 ·········· 19
  3　Property/ 性能 ·········· 22
  4　Scouring Process/ 煮练工艺 ·········· 26
    4.1　Evaluation of Pectin Content/ 果胶含量测定 ·········· 26
    4.2　Alkali Scouring/ 碱煮练 ·········· 27
    4.3　Pectase Scouring/ 果胶酶煮练 ·········· 28
    4.4　Comparisons of Alkali and Pectase Scouring/
        碱煮练和果胶酶煮练对比 ·········· 29

  4.5 Color Fastness of Natural Colored Cotton Fabrics After Alkali and Pectase Scouring/ 碱煮练和果胶酶煮练的天然彩棉织物的色牢度 ········· 30

 5 Fiber Identification/ 鉴别 ················· 31
  5.1 Washing Method/ 洗涤法 ············· 31
  5.2 Sectioning Method/ 切片法 ············ 31

 6 Application/ 应用 ······················ 32
 7 Problem/ 问题 ························ 32
 Exercises/ 练习 ·························· 33
 References/ 参考文献 ······················ 34

Chapter Three Bamboo Fiber/ 竹纤维 ················ 35

 1 Introduction/ 前言 ····················· 35
  1.1 Bamboo Application/ 竹子用途 ·········· 35
  1.2 Bamboo Structure/ 竹子结构 ············ 36
  1.3 Fiber Distribution in Bamboo/ 竹子中纤维分布 ······ 37
  1.4 Fiber Length and Fineness in Bamboo/ 竹子中纤维长度和细度 ············ 37

 2 Preparation/ 制备 ····················· 38
  2.1 Preparation of Natural Bamboo Fiber/ 竹原纤维制备 ··························· 38
  2.2 Preparation of Bamboo Pulp Fiber/ 竹浆纤维制备 ··························· 39

 3 Structure/ 结构 ······················· 40
  3.1 Molecular Structure/ 分子结构 ··········· 40
  3.2 Supramolecular Structure/ 超分子结构 ······ 42
  3.3 Morphological Structure/ 形态结构 ········ 44

 4 Property/ 性能 ······················· 45
 5 Application/ 应用 ····················· 49

6　Problem/ 问题 ·· 50
　　Exercises/ 练习 ·· 51
　　References/ 参考文献 ·· 51

Part Two　Protein Fiber/ 蛋白质纤维

Chapter Four　Milk Protein Fiber/ 牛奶蛋白纤维 ············ 54
　　1　Introduction/ 前言 ·· 54
　　2　Preparation/ 制备 ·· 55
　　　2.1　Principle/ 原理 ······································ 55
　　　2.2　Preparation Technology/ 制备工艺 ················ 56
　　3　Structure/ 结构 ·· 60
　　　3.1　Chemical Composition/ 化学组成 ·················· 60
　　　3.2　Morphological Structure/ 形态结构 ················ 60
　　4　Property/ 性能 ·· 61
　　　4.1　Tensile Property/ 强伸性 ··························· 61
　　　4.2　Moisture Absorption and Conductivity/
　　　　　吸湿导湿性 ·· 62
　　　4.3　Friction Property/ 摩擦性 ·························· 62
　　　4.4　Electrical Property/ 电学性 ························ 63
　　　4.5　Dyeability/ 染色性 ·································· 63
　　　4.6　Comfort/ 舒适性 ···································· 63
　　　4.7　Aesthetics/ 外观 ···································· 64
　　　4.8　Light Resistance/ 耐光性 ·························· 64
　　　4.9　Healthcare Function/ 保健功能 ···················· 64
　　5　Fiber Identification/ 纤维鉴别 ·························· 65
　　6　Problem/ 问题 ·· 65
　　　6.1　Poor Heat Resistance/ 耐热性差 ·················· 65

  6.2 Low Chemical Stability/ 化学稳定性低 ·················· 65
  6.3 Color/ 颜色 ·················· 66
  6.4 High Price/ 价格高 ·················· 66
 7 Application/ 应用 ·················· 66
  7.1 Yarns/ 纱线类 ·················· 66
  7.2 Woven and Knitted Fabrics/ 机织和针织面料 ·················· 66
  7.3 Home Textiles/ 家纺类 ·················· 67
  7.4 Nonwoven Fabrics/ 非织造布 ·················· 67
 Exercises/ 练习 ·················· 68
 References/ 参考文献 ·················· 68

Chapter Five Soybean Protein Fiber/ 大豆蛋白纤维 ·················· 69
 1 Introduction/ 前言 ·················· 69
 2 Preparation/ 制备 ·················· 70
 3 Structure/ 结构 ·················· 71
  3.1 Chemical Structure/ 化学结构 ·················· 71
  3.2 Morphological Structure/ 形态结构 ·················· 71
 4 Property/ 性能 ·················· 72
  4.1 Mechanical Property/ 机械性能 ·················· 72
  4.2 Crimp Property/ 卷曲性 ·················· 74
  4.3 Antistatic Property/ 抗静电性 ·················· 75
  4.4 Optical Property/ 光学性能 ·················· 75
  4.5 Heat Resistance/ 耐热性 ·················· 76
  4.6 Appearance and Handle/ 外观与手感 ·················· 76
  4.7 Moisture Absorption and Fast Drying/
    吸湿快干性 ·················· 76
  4.8 Dyeability/ 染色性 ·················· 77
  4.9 Antibacterial Activity/ 抗菌性 ·················· 77
  4.10 Healthcare Function/ 保健功能 ·················· 77

  4.11 Comfort/ 舒适性 ·············· 77
  4.12 Chemical Resistance/ 耐化学性 ·············· 78
 5 Application/ 应用 ·············· 78
Exercises/ 练习 ·············· 79
References/ 参考文献 ·············· 80

Chapter Six Modified Wool Fiber/ 改性羊毛纤维 ·············· 81
 1 Introduction/ 前言 ·············· 81
 2 Surface Modified Wool Fiber/ 表面改性羊毛纤维 ·············· 82
  2.1 Oxidation Treatment/ 氧化法 ·············· 83
  2.2 Enzyme Treatment/ 酶处理 ·············· 85
  2.3 Plasma Treatment/ 等离子体处理 ·············· 86
  2.4 Polymer Treatment/ 聚合物处理 ·············· 89
 3 Slenderized Wool Fiber/ 细化羊毛纤维 ·············· 90
  3.1 Decrement Slenderization (Mercerization)
    Treatment/ 减量细化（丝光）法 ·············· 90
  3.2 Stretching Slenderization Technology (SST)/
    拉伸细化技术 ·············· 90
 4 Bulking Wool Fiber/ 膨化羊毛纤维 ·············· 94
 5 Colored Wool Fiber/ 彩色羊毛纤维 ·············· 96
Exercises/ 练习 ·············· 96
References/ 参考文献 ·············· 96

Part Three Synthetic Fiber/ 合成纤维

Chapter Seven Polytrimethylene Terephthalate (PTT) Fiber/
    聚对苯二甲酸丙二醇酯纤维 ·············· 100
 1 Introduction/ 前言 ·············· 100
 2 Manufacturing/ 制备 ·············· 101

  2.1 Preparation of 1, 3-propanediol (PDO)/
    1, 3- 丙二醇制备 ·· 101
  2.2 PTT Polymerization/PTT 合成 ································· 103
  2.3 Spinning/ 纺丝 ·· 107
 3 Structure/ 结构 ··· 107
 4 Property/ 性能 ·· 108
 5 Application/ 应用 ··· 109
 Exercises/ 练习 ·· 110
 References/ 参考文献 ··· 110

Part Four High Performance Fiber/ 高性能纤维

Chapter Eight Carbon Fiber/ 碳纤维 ··································· 112
 1 Introduction/ 引言 ·· 112
 2 Preparation/ 制备 ··· 114
 3 Structure/ 结构 ··· 116
  3.1 Sheath-Core Structure Model/ 皮芯结构模型 ·········· 116
  3.2 Microfibril Model/ 微原纤模型 ······························· 117
 4 Property/ 性能 ·· 117
  4.1 Mechanical Property/ 力学性能 ······························ 117
  4.2 Thermal Property/ 热学性能 ··································· 119
  4.3 Other Properties/ 其他性能 ···································· 120
 5 Application/ 应用 ··· 120
 Exercises/ 练习 ·· 121
 References/ 参考文献 ··· 122

Chapter Nine Glass Fiber/ 玻璃纤维 ··································· 123
 1 Introduction/ 前言 ·· 123
  1.1 GF Development/ 玻璃纤维发展概况 ····················· 123

  1.2 GF Classification/ 玻璃纤维分类 ·········· 125
 2 Preparation/ 制备 ·········· 127
 3 Structure/ 结构 ·········· 128
 4 Property/ 性能 ·········· 129
  4.1 Appearance and Density/ 外观和密度 ·········· 129
  4.2 Mechanical Property/ 力学性能 ·········· 130
  4.3 Thermal Property/ 热学性能 ·········· 132
  4.4 Electrical Property/ 电学性能 ·········· 133
  4.5 Chemical Property/ 化学性能 ·········· 134
 5 Application/ 应用 ·········· 135
 Exercises/ 练习 ·········· 136
 References/ 参考文献 ·········· 137

Chapter Ten Aromatic Polyamide Fiber/ 芳香族聚酰胺纤维 ·········· 138
 1 Introduction/ 前言 ·········· 138
 2 Preparation/ 制备 ·········· 139
  2.1 Polymer synthesis/ 高聚物合成 ·········· 139
  2.2 Spinning/ 纺丝 ·········· 141
 3 Structure/ 结构 ·········· 142
 4 Property/ 性能 ·········· 143
  4.1 Mechanical property/ 力学性能 ·········· 143
  4.2 Thermal property/ 热学性能 ·········· 144
 5 Application/ 应用 ·········· 145
 Exercises/ 练习 ·········· 148
 References/ 参考文献 ·········· 149

Chapter Eleven Ultra-high Molecular Weight Polyethylene Fiber/ 超高分子量聚乙烯纤维 ·········· 150
 1 Introduction/ 前言 ·········· 150
 2 Manufacturing/ 制备 ·········· 152

2.1 Spinning Solution/纺丝液 ········· 153
  2.2 Gelation and Crystallization/凝胶化和结晶 ········· 155
  2.3 Extraction/萃取 ········· 155
  2.4 Drawing/牵伸 ········· 156
3 Structure/结构 ········· 157
4 Property/性能 ········· 158
  4.1 Density/密度 ········· 159
  4.2 Tensile Property/拉伸性能 ········· 159
  4.3 Mechanical Property in Transverse Direction/横向力学性能 ········· 161
  4.4 Viscoelasticity/黏弹性能 ········· 161
  4.5 Impact Resistance and Bulletproof/耐冲击性能和防弹性能 ········· 162
  4.6 Fatigue and Abrasion Resistance/疲劳和耐磨性能 ········· 162
  4.7 Hydrophobicity/拒水性能 ········· 163
  4.8 Chemical Resistance/耐化学性能 ········· 163
  4.9 Resistance to Light and Other Radiation/耐光和其他辐射性能 ········· 165
  4.10 Electrical Property/电学性能 ········· 165
  4.11 Acoustic Property/声学性能 ········· 165
  4.12 Biological Resistance and Toxicity/耐生物性能和毒性 ········· 166
  4.13 Thermal Resistance/耐热性能 ········· 166
  4.14 Shrinkage/收缩性能 ········· 166
5 Application/应用 ········· 168
  5.1 National Defense/国防 ········· 168
  5.2 Aerospace/航空航天 ········· 169

  5.3 Industrial and Civil/工业用和民用 ················ 169
  Exercises/练习 ···································· 171
  References/参考文献 ······························ 171

Words and Terminologies/生词和专业词汇

Chapter One Kapok ···································· 172
Chapter Two Natural Colored Cotton Fiber ············ 176
Chapter Three Bamboo Fiber ························· 178
Chapter Four Milk Protein Fiber ····················· 181
Chapter Five Soybean Protein Fiber ·················· 185
Chapter Six Modified Wool Fiber ····················· 187
Chapter Seven PTT Fiber ···························· 190
Chapter Eight Carbon Fiber ························· 192
Chapter Nine Glass Fiber ····························· 194
Chapter Ten Aromatic Polyamide Fiber ················ 196
Chapter Eleven UHMWPE Fiber ······················ 199

Part One

Cellulose Fiber/ 纤维素纤维

Chapter One   Kapok Fiber/ 木棉纤维

As a natural **cellulose** fiber, **kapok** fiber shares many similarities with cotton fiber. In contrast to cotton fiber, kapok fiber has many unique advantages in **density, luster, sound absorption, warmth retention,** etc.

木棉纤维是一种天然纤维素纤维，与棉纤维有许多相似之处，但在密度、光泽、吸声性和保暖性等方面具有独特优势。

1   Introduction/ 前言

Kapok fiber comes from kapok tree. Kapok tree belongs to the family of **bombacaceae** and primarily grows in Asia, Africa and South America. It is drought-tolerant and not cold-resistant. It is suitable for growing in warm, dry and sunny environment. Both kapok fiber and cotton fiber belong to single-cell fiber. Cotton fiber is **seed fiber** and derived from the **epidermal cell** which is firmly attached to the seed. Kapok fiber is **fruit fiber** and derived from the **lining cell** which is attached to the inner wall of the **capsule shell**. And kapok fiber can be easily separated from the shell due to the low adhesion. The preliminary processing of kapok fiber therefore is relatively convenient compared with that of cotton fiber. While the **cotton ginning** is necessary for cotton, the kapok seed only needs to be picked out by hand. It also can be put into bamboo **crate** and **sieve**d, and the kapok seed itself will sink to the bottom.

There are basically two ways to classify kapok fiber, by color and by **species**. According to color, it can be divided into three categories: white, light yellow and light brown. According to species, it can be divided into six categories: bombax malabaricum, beiba mill, ochroma swartz, durio adans, pachira aubl and adansonia linn. Kapok fiber from

木棉纤维来自木棉树，木棉树属于木棉科植物，主要生长在亚洲、非洲和南美洲，耐旱不耐寒，宜种植在温暖干燥和阳光充足的环境中。木棉与棉纤维均属单细胞纤维。棉纤维是种子纤维，由种子的表皮细胞生长而成，附着于种子上；而木棉纤维是果实纤维，由内壁细胞生长而成，附着于木棉蒴果壳体内壁。木棉纤维在蒴果壳体内壁的附着力小，容易分离。因此，木棉纤维的初步加工比较方便，不像棉花那样须经过轧棉加工，只要手工将木棉种子剔出或装入箩筐中筛动，木棉种子即自行沉底。

木棉纤维可按品种和颜色分类，根据颜色，木棉纤维有白色、淡黄色和浅棕色3种；根据品种，木棉纤维有木棉属、吉贝属、轻木属、榴莲属、瓜栗属和猴面包树属6种。中国产木棉纤

China mainly belongs to bombax malabaricum.

维主要属于木棉属。

2　Structure/ 结构

2.1　Molecular Structure/ 分子结构

The molecular structural parameters of kapok and cotton fibers are shown in Table 1–1. Kapok fiber is primarily composed of 35%~65% cellulose, 13%~22% **lignin**, 22%~45% **hemicellulose** and 10%~11% non**cellulosic** substances such as wax, ash. About 94%~95% of the cotton fiber is cellulose, 5%~6% is hemicellulose, and 7%~8% is noncellulosic substances. The **degree of polymerization** in kapok fiber is only 1000, much lower than that in cotton fiber (6000~15000).

木棉和棉纤维的分子结构参数见表1-1。木棉纤维主要由纤维素、木质素、半纤维素和蜡质、灰分等其他非纤维素物质组成，其含量分别为35%~65%、13%~22%、22%~45%和10%~11%。棉纤维主要由纤维素、半纤维素和非纤维素物质组成，其含量分别为94%~95%、5%~6%和7%~8%。木棉纤维的聚合度只有1000，远小于棉纤维的聚合度（6000~15000）。

Table 1–1　Molecular structural parameters of kapok and cotton fibers
表1-1　木棉和棉纤维分子结构参数

Fiber 纤维	Chemical composition 化学组成（%）				Polymerization degree 聚合度
	Cellulose 纤维素	Lignin 木质素	Hemicellulose 半纤维素	Noncellulosic substance 非纤维素物质	
Kapok 木棉纤维	35~65	13~22	22~45	10~11	1000
Cotton 棉纤维	94~95	0	5~6	7~8	6000~15000

As shown in Table 1–1, there are significant differences in the chemical composition between kapok and cotton fibers. Kapok fiber contains a certain amount of lignin while no lignin ever exists in cotton fiber. The cellulose content in kapok fiber is much lower than that in cotton fiber, while the hemicellulose content in kapok fiber

由表1-1可知，木棉和棉纤维的化学组成及其含量有较大差异。木棉纤维中有木质素，而棉纤维中没有木质素。木棉纤维的纤维素含量远低于棉纤维，而半纤维素含量高于棉纤维。

is much higher than that in cotton fiber.

2.2 Supramolecular Structure/ 超分子结构

Kapok fiber is composed of two layers with different **microfibrillar orientation**s. The outer layer is composed of cellulose microfibrils **perpendicular** to the fiber axis, whereas the inner layer is composed of microfibrils **parallel** to the fiber axis.

The degree of **crystallization** in kapok, cotton and **flax** fibers is shown in Table 1-2. As shown below, the degree of crystallization in kapok fiber is 35.9%, much lower than it is in cotton (65%~72%) and in flax fibers (69%).

木棉纤维由两层结构组成，其微原纤取向度不同。外层由垂直于纤维轴的纤维素微原纤组成，内层由平行于纤维轴的微原纤组成。

木棉、棉和亚麻纤维的结晶度见表1-2，分别为35.9%、65%~72%和69%，木棉纤维的结晶度低于棉纤维和亚麻纤维。

Table 1-2 Crystallization degree of kapok, cotton and flax fibers
表 1-2 木棉、棉和亚麻纤维的结晶度

Fiber 纤维	Kapok 木棉	Cotton 棉	Hemp 亚麻
Crystallization degree 结晶度（%）	35.9	65~72	69

2.3 Morphological Structure/ 形态结构

Figure 1-1 shows the SEM images of kapok and cotton fibers. As shown in Figure 1-1(a) and Figure 1-1(b), cotton fiber has a **flat ribbon longitudinal** form with **natural convolution** and a **kidney-shaped cross section** with small **lumen**. The lumen is once full of **cell sap**. As the sap evaporates, the fiber will collapse inwards accounting for the peculiar **morphology**. The hollow degree in cotton fiber is 10%. As shown in Figure 1-1(c) and Figure 1-1(d), kapok fiber appears as a cylindrical shape, a smooth surface without natural convolution in direction along the fiber. The fiber has a round root, **coarse** central portion and **fine** tip, and both ends are closed as shown in Figure 1-1(c). The cross section of kapok fiber is **elliptical** to nearly circular with **thin** fiber wall and large lumen as shown in Figure

木棉和棉纤维形态结构的电子扫描显微镜照片如图1-1所示。由图1-1（a）和图1-1（b）可以看出，棉纤维纵向呈扁平带状，有天然转曲，截面呈腰圆形，有小中腔，中腔内一度充满了细胞液，随着细胞液的蒸发，棉纤维向内塌陷，形成了其独特的形态，中空度约为10%。由图1-1（c）和图1-1（d）可以看出，木棉纤维纵向呈圆柱型，表面光滑，无天然转曲，纤维根部圆、中段粗、梢端细、两段封闭［图1-1（c）］，截面呈椭圆形或圆

1-1(d). The fiber wall is only 0.5~2.0 μm and appears almost transparent. The ratio of fiber width to wall **thickness** is about 20 and the fiber is easily crushed. The diameter of lumen is 10~15 μm when it is full of air, which is 10 times that of the fiber wall. It has been estimated that kapok fiber has a **hollow degree** of over 80%~90%, which is 2~3 times that of cotton fiber as shown in Figure 1-1(e). In current, hollow degree of chemical fibers is only as much as 40%. Hence, kapok fiber is the top with the highest hollow degree. Kapok fiber takes on balloon-like and ribbon-like morphology before and after the breakdown of the cell in the cross section as shown in Figure 1-1(f).

形，壁薄，有大中腔［图1-1（d）］，纤维壁厚仅为0.5~2 μm，接近透明，纤维宽与纤维壁厚比值约为20，纤维容易被压扁，中腔充满空气时，直径为10~15 μm，是壁的10倍，中空度高达80%~90%［图1-1（e）］，是棉纤维的2~3倍。目前化学纤维的中空度最高达40%，因此，木棉纤维是中空度最高的纤维。截面细胞未破裂时纤维呈气囊状，破裂后呈扁平带状［图1-1（f）］。

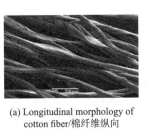
(a) Longitudinal morphology of cotton fiber/棉纤维纵向

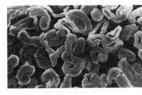
(b) Cross sectional morphology of cotton fiber/棉纤维横向

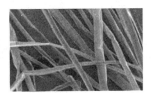
(c) Longitudinal morphology of kapok fiber/木棉纤维纵向

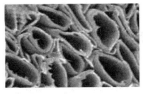
(d) Cross sectional morphology of kapok fiber/木棉纤维横向

(e) Thin wall and large lumen of kapok fiber/木棉纤维薄壁和大中腔

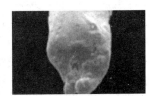
(f) Kapok fiber root/木棉纤维根端

Figure 1-1　SEM images of morphology of kapok and cotton fibers
图1-1　木棉和棉纤维形态的扫描电子显微镜照片

In general, though cotton and kapok fibers belong to single-cell fiber, there exist **distinct** differences in the fiber morphological structure. Cotton fiber appears as a flat, twisted ribbon and its structure is more compact with small lumen. And kapok fiber presents as a cylinder and its structure is looser with large lumen.

总的来说，虽然木棉与棉纤维同属单细胞纤维，但是在纤维形态结构上存在着较大的差异。棉纤维呈扁平转曲带状，中腔小，结构较紧密；木棉纤维呈圆柱形，中腔大，结构较松散。

3 Property/ 性能

The unique fibrous structure in kapok fiber differentiates its physical and chemical properties from those of cotton fiber.

由于独特的纤维结构,木棉纤维与棉纤维具有不同的物理和化学性能。

3.1 Physical Property/ 物理性能

The basic physical properties of kapok and cotton fibers are summarized in Table 1–3.

木棉与棉纤维的基本物理性能见表 1–3。

Table 1–3　Physical properties of kapok and cotton fibers
表 1–3　木棉和棉纤维的物理性能

Property 性能	Kapok 木棉	Cotton 棉
Length 长度（mm）	8~32	23~45
Linear density 线密度（dtex）	0.9~1.2	1.65~1.95
Density 密度（g/cm^3）	0.29	1.54~1.55
Breaking tenacity 断裂强度（cN/dtex）	0.8~1.4	2.0~2.4
Elongation ratio at break 断裂伸长率（%）	1.5~3.0	7~9
Initial modulus 初始模量（cN/dtex）	48.7~114.5	60~82
Torsional rigidity 扭转刚度（cN·cm^2/tex^2）	71.5 × 10^{-4}	—
Compression modulus 压缩模量（kPa）	43.63	—
Moisture regain 回潮率（%）	10~10.73	8.5
Dyeing ratio 上染率（%）	63	88
Refractive index 光折射率（%）	1.718	1.596

Length and fineness: As shown in Table 1-3, kapok fiber belongs to **staple fiber**, ranging from 8 mm to 32 mm in length, and is shorter than the cotton fiber which ranges from 23 mm to 45 mm in length. Among textile fibers, kapok fiber is the finest natural cellulose fiber so far with the average diameter of 30~36 μm. Kapok fiber has a small linear density ranging from 0.9 dtex to 1.2 dtex, approximately half the cotton fiber (1.65~1.95 dtex).

Density: Due to larger lumen and thinner wall, kapok fiber has a remarkably lower density (0.29 g/cm^3) than cotton fiber (1.54~1.55 g/cm^3). Kapok fiber is for now the lightest natural cellulose fiber.

Tensile property: Compared with cotton fiber, kapok fiber contains a large amount of lignin, which improves torsional rigidity and lowers breaking tenacity, elongation ratio at break, initial modulus as well as compression modulus. Kapok fiber has a higher torsional rigidity (71.5 × 10^{-4} cN·cm^2/tex^2) than glass fiber, which decreases **twisting** efficiency. The breaking tenacity, breaking elongation ratio, initial modulus of kapok fiber are 0.8~1.4 cN/dtex, 1.5%~3.0%, 48.7~114.5 cN/dtex, respectively, which are all lower than those of cotton fiber. The compression modulus of kapok fiber is only 43.63 kPa so that it could easily collapse inwards under high pressure. The lumen loss of dry kapok fiber is less than that of wet kapok fiber under the same pressure, so is the **plastic deformation**. Therefore, it is appropriate to reduce environmental relative humidity to maintain the hollow structure in kapok fiber during textile processing.

Compression modulus: The compression modulus of kapok fiber is low. It can bear low pressure and is easily to be deformed or even be flattened. Therefore, the high

长度和细度：由表1-3可知，木棉纤维属于短纤维，其长度为8~32 mm，比棉纤维（23~45 mm）短。纺织纤维中，木棉纤维是目前最细的天然纤维素纤维，平均直径为30~36 μm。木棉纤维的线密度为0.9~1.2 dtex，约为棉纤维（1.65~1.95 dtex）的一半。

密度：由于木棉纤维较大的中腔和较薄的壁，木棉纤维的密度为0.29 g/cm^3，远小于棉纤维（1.54~1.55 g/cm^3）。木棉纤维是目前最轻的天然纤维素纤维。

强伸性：与棉纤维相比，木质素的存在提高了木棉纤维的扭转刚度，降低了木棉断裂强度、断裂伸长率、初始模量和压缩模量。纤维的扭转刚度为71.5 × 10^{-4} cN·cm^2/tex^2，大于玻璃纤维，降低加捻效率。断裂强度、断裂伸长率和初始模量分别为0.8~1.4 cN/dtex，1.5%~3.0%和48.7~114.5 cN/dtex，均小于棉纤维。木棉纤维的压缩模量仅为43.63 kPa，当压力过大时，纤维中腔内的空气会被挤出，极易被压扁。在受到相同压力时，干燥木棉纤维的中腔损失小于潮湿木棉纤维，塑性变形程度轻。因此，在纺织加工时，可以适当降低环境相对湿度来保持木棉纤维的中腔结构。

压缩模量：木棉纤维压缩模量小，承受较小的压力，受压容易变形，甚至被压扁，因此，木

pressure should be avoided during processing and using. The air in its lumen is easily to be squeezed out, resulting in irreversible flattening. Many excellent properties of kapok fiber will be lost.

Moisture absorption: The pretreated kapok fiber has a moisture regain of about 10%~10.73%, higher than cotton fiber (8.5%), but close to **mercerize**d cotton fiber (10.6%). Several aspects of kapok fiber account for its moisture regain. Firstly, kapok fiber contains a large number of noncellulosic substances which are **hydrophilic** and can quickly absorb water. Secondly, kapok fiber has lower **crystallinity** than cotton fiber. Water molecules are more likely to enter **amorphous** regions. Thirdly, the lignin and flowing air in kapok fiber help to accelerate water transport in fiber. In addition, the special hollow structure increases fiber **specific surface area**. All above these are beneficial to excellent moisture absorption of kapok fiber.

Warmth retention: Kapok fiber has high hollow degree and closed lumen structure, which can not be achieved by existing **filament spin**ning technology. The larger lumen in kapok fiber is able to contain much more **static** air, resulting in lower **thermal conductivity**. Therefore, kapok fiber shows excellent warmth retention.

Luster: Kapok fiber exhibits good luster. It could be explained by following two reasons. Firstly, there is no natural convolution in direction along kapok fiber length and the shape of cross section is close to round. Secondly, the surface of kapok fiber is covered by a layer of wax. Accordingly, kapok fiber has a higher average refractive index (1.718) than cotton fiber (1.596).

Dyeability: Kapok fiber contains large amounts of cellulose and lignin which show high **affinity** to dyes in aqueous solution. However, the larger lumen and

棉纤维在加工和使用过程中，应该注意避免高压。在高压下，木棉纤维中腔中的空气被压出，造成不可恢复的压扁，损失木棉纤维的许多优良性能。

吸湿性：经过前处理的木棉纤维的回潮率约为10%~10.73%，高于棉纤维（8.5%），接近于丝光棉纤维（10.6%）。原因如下：第一，由于木棉纤维的非纤维素物质含量较高，亲水性好，能快速地吸水。第二，木棉纤维的结晶度低于棉纤维，水分子更容易进入无定形区。第三，大量木质素的存在以及内部空气的流动，加快水分在纤维内部传输。此外，纤维独特的中空结构增加纤维比表面积。这些都会使木棉纤维具有优异的吸湿性。

保暖性：木棉纤维具有现有纺丝技术无法达到的高中空度和封闭的中腔结构。木棉纤维较大的中腔能容纳较多的静止空气，使得木棉纤维的导热系数较小，因此，木棉纤维具有优异的保暖性。

光泽：木棉纤维光泽较好，主要有两方面原因。其一，纤维纵向无天然转曲，横截面接近圆形；其二，木棉纤维表面覆盖一层蜡质。因此，木棉纤维的平均光折射率为1.718，高于棉纤维（1.596）。

染色性：木棉纤维含有大量亲水的纤维素和木质素，与溶液中染料之间亲和力较强，但是，

waxy surface are not favorable for the access of dyes. A comparatively low dyeing ratio is observed for kapok fiber, which is only 63%. The dyeing ratio for cotton fiber is 88% under the same condition.

Hydrophobicity oleophilicity: Kapok fiber displays excellent **hydrophobic–oleophilic** behavior due to the wax layer on its surface.

Sound insulation and absorption: The larger lumen endows kapok fiber with good sound **insulation** and absorption.

Spinnability: The **cohesive force** among fibers is reduced due to shorter length as well as the wax on the surface. Meanwhile, it is difficult to twist kapok fiber due to higher torsional rigidity. The spun yarn is **thick**er, lower in tenacity and elongation ratio, poorer in **evenness** and has more **hairiness** on the surface. Hence, the spinnability of kapok fiber is poor and it is not suitable to be purely spun into yarn. However, it could be successfully blended spun with other fibers to improve its spinnability.

Compared with cotton yarn, kapok/cotton **blended yarn** presents cotton-like **moisture permeability**, lower **air permeability**, and better warmth retention as well as **moisture conductivity** owing to the highly **porous** structure in kapok fiber. The warmth retention and the dyeability of kapok/cotton blended yarn become better and worse with the increasing content of kapok fiber, respectively.

3.2 Chemical Property/ 化学性能

Acid/alkali resistance: Kapok fiber has good acid and alkali resistance. Table 1-4 gives the **solubility** of kapok fiber in different **acidic** and **alkaline solvent**s at a certain temperatwre for a given time.

As shown in Table 1-4, kapok fiber does not

较大的中腔和纤维表面的蜡质层不利于染料的进入，使得木棉纤维的上染率较低，仅为63%，而同样条件下，棉纤维的上染率为88%。

疏水亲油性：木棉纤维表面的蜡质使其具有极好的疏水亲油性。

隔音吸声性：木棉纤维较大的中腔使其具有极好的隔音吸声性。

可纺性：木棉纤维较短且表面存在蜡质降低了纤维之间的抱合力，同时较高的扭转刚度，导致加捻困难，所纺纱线较粗、强度和伸长率较低、条干较差以及纱线表面毛羽较多。因此，木棉纤维可纺性差，不适合纯纺，但是，可以与其他纤维混纺，改善其可纺性。

与棉纱相比，由于木棉纤维的多孔结构，木棉/棉混纺纱具有与棉类似的透湿性、较低的透气性、较好的保温性和导湿性。随着木棉纤维含量的增加，混纺纱的保温性变好，但染色性变差。

耐酸碱性：木棉纤维具有良好的耐酸碱性，在一定温度时，木棉纤维在不同酸性和碱性溶剂中一定时间后的溶解性见表1-4。

由表1-4可知，常温下木

dissolve at room temperature in the solution such as 20% **hydrochloric acid** (HCl), 53% **nitric acid** (HNO$_3$), 100% **acetic acid** (CH$_3$COOH), 80% **formic acid** (HCOOH) and 5% **sodium hydroxide** (NaOH). It partially dissolves in 60% **sulphuric acid** (H$_2$SO$_4$) at 50 ℃ and 30% HCl at 100 ℃, and completely dissolves in 75% H$_2$SO$_4$ at 30 ℃ and 65% HNO$_3$ at 100 ℃. Kapok fiber is resistant to **dilute** strong acid, weak acid and dilute strong alkaline at room temperature but not resistant to **concentrated** strong acid. The acid and alkali resistance of kapok fiber is affected by solvent **concentration**, **dissolution** temperature and dissolution time. The solubility of kapok fiber in strong concentrated acid and alkaline solution rises with the increasing solvent concentration, and dissolution time and temperature.

棉纤维在20%盐酸（HCl）、53%硝酸（HNO$_3$）、100%乙酸（CH$_3$COOH）和80%甲酸（HCOOH）和5%氢氧化钠（NaOH）中不溶解。在50℃、60%硫酸（H$_2$SO$_4$）和100℃、30% HCl中部分溶解。在30℃、75% H$_2$SO$_4$和100℃、65% HNO$_3$完全溶解。常温下木棉纤维耐稀强酸、弱酸和稀强碱，但不耐浓强酸。溶剂浓度、溶解温度和时间影响木棉纤维的耐酸碱性。随溶剂浓度、溶解温度和时间增加，木棉纤维在高浓度强酸和强碱中的溶解度增加。

Table 1-4　Acid and alkali resistance of kapok fiber
表1-4　木棉纤维的耐酸碱性

Solvent 试剂	Dissolution temperature 溶解温度（℃）	Dissolution time 溶解时间（min）	Result 结果
5% Sodium hydroxide 5%氢氧化钠	Room temperature 室温	20	**Insoluble** 不溶解
60% Sulphuric acid 60%硫酸	50	15	Partially **soluble** 部分溶解
75% Sulphuric acid 75%硫酸	30	30	Completely soluble 全部溶解
20% Hydrochloric acid 20%盐酸	Room temperature 室温	15	Insoluble 不溶解
30% Hydrochloric acid 30%盐酸	100	20	Partially soluble 部分溶解
53% Nitric acid 53%硝酸	Room temperature 室温	30	Insoluble 不溶解
65% Nitric acid 65%硝酸	100	15	Completely soluble 全部溶解
100% Acetic acid 100%乙酸	Room temperature 室温	15	Insoluble 不溶解
80% Formic acid 80%甲酸	Room temperature 室温	20	Insoluble 不溶解

Thermal stability: Thermal **degradation** temperature is the most important indicator for the thermal stability of fiber. Low crystallinity and high hemicellulose content in kapok fiber provide relatively lower thermal degradation temperature than that in cotton fiber. Consequently, the thermal stability of kapok fiber is poorer than that of cotton fiber. However, it requires a long time for kapok fiber to achieve maximum degradation rate and kapok fiber is thermally **degrade**d in a large span of temperature and time. Therefore, the **thermal resistance** of kapok fiber is better than that of cotton fiber.

Antibacterial property: Kapok fiber has a killing effect on **mould** and a **broad spectrum antimicrobial** effect. There are two main reasons. For fiber structure, owing to its thin wall, high hollow degree and large specific surface area, kapok fiber is rich in oxygen, which is considered to hinder the **reproduction** of **anaerobic bacteria** and **mite** on the fiber surface. For chemical composition, there are naturally antimicrobial substances in fiber such as **flavonoid** and **triterpen**e. Kapok fiber performs better antibacterial activity against **gram-positive escherichia coli** (E. coli) than against **gram-negative staphylococcus aureu** (S. aureu).

热稳定性：热降解温度是衡量纤维热稳定性的重要指标。木棉纤维的结晶度低于棉纤维，半纤维素含量高于棉纤维，热降解温度低于棉纤维，热稳定性比棉纤维差。但是，木棉纤维达到最大降解速率需要的时间较长，其发生热降解的温度和时间区间大。所以，木棉纤维的耐热性好于棉纤维。

抗菌性：木棉纤维对霉菌具有杀灭作用和广谱抗菌的效果。主要原因有两方面：从纤维结构上来说，木棉纤维壁薄、中空度高、比表面积大，纤维中富含氧气，阻碍厌氧菌和螨在纤维表面的繁殖；从化学组成上来说，纤维中含有黄酮类和三萜类天然抗菌物质。木棉纤维对革兰氏阳性的大肠杆菌的抗菌性好于对革兰氏阴性的金黄色葡萄球菌。

4 Application/ 应用

The unique features of kapok fiber are determined by its special structure. It has many advantages of lightness, **softness**, smoothness, luster, moisture absorption, warmth retention, hydrophobicity, oleophilicity, sound insulation and sound absorption and so on. Therefore, kapok fiber and its **composite** have been gaining increasing attention in recent years. Kapok fiber can be directly used as **filler** and **reinforcing material**, or blended spun with other cellulosic fibers which ultimately to be applied as **clothing** and **home textile**.

特殊的结构决定了木棉纤维独特的性能，木棉纤维具有轻质、柔软、光滑、吸湿、保温、拒水吸油、隔声吸音、有光泽等优点，因此，近年来，木棉纤维及其复合材料越来越受到重视。木棉纤维可以直接用作填充、增强等材料或与其他纤维混纺，制织服装、家纺等面料。

Apparel and home textile: The pretreated kapok fiber can be blended spun with cotton, **viscose** or other cellulosic fibers to be used as apparel and home textile with good luster and **handle** due to its softness, warmth retention, moisture absorption and air permeability.

Stuffing material: Kapok fiber is applied as stuffing materials for **bedding** and **upholstery** such as **pillow**, quilt, **mattress**, **cushion** and so on owing to its low density, good warm retention and **antibacterial activity**.

Buoyant material: Kapok fiber is used as **life preserver** and other water-safety equipment because of its low density and hydrophobicity. Kapok fiber can bear the weight 20~30 times of self-weight and float on water for hours without absorbing water and sinking, thus it is used as an excellent buoyant material. Life jacket with kapok fiber as stuffing material will not age in use. Nevertheless, the **buoyancy** of life jacket or other buoyant material will drop after being used for a long time since the lumen is easily broken and the fiber is easily squashed due to its high hollow degree. The optimum density of kapok fiber based buoyant material is about 0.036~0.050 g/cm^3. In addition, treatment processes such as **lamination** laying **wadding** and hot-**melt** bonding are applied to improve the **compression resistance** of kapok **fiber assembly**, which could better meet the requirements of buoyant material.

Oil absorbing and water repellent material: Kapok fiber is used as oil absorbing material and water repellent packaging paper as a result of its excellent hydrophobicity-oleophilicity.

Heat insulation and sound absorbing material: Kapok fiber is considered to be used as stuffing materials for the **heat insulation** layer and sound absorption layer of house due to its warmth retention and sound insulation. The heat absorption and heat retention of kapok/wool heat insulation composite are better those of wool heat insulation material.

服装和家纺面料：由于其柔软、保温、吸湿和透气，经过预处理的木棉纤维可以与棉、黏胶纤维或其他纤维素纤维混纺，制备光泽和手感良好的服装和家用纺织品。

填充材料：由于其密度小、保温和抗菌，木棉纤维用于枕芯、被子、褥垫、靠垫等床品和室内装饰品的填充材料。

浮力材料：由于其密度小、拒水，木棉纤维用于救生用具和其他水上安全设备。木棉纤维在水中可承受自重的20~30倍，且数小时不吸水不下沉，是很好的浮力材料。木棉纤维为填充材料做成的救生衣在穿着使用过程中不会产生老化。然而，由于中空度高，中腔易破裂、纤维易被压扁，救生衣或其他浮力材料长期使用后，浮力会下降。木棉纤维浮力材料的最佳密度为0.036~0.050 g/cm^3。另外，可通过分层铺絮以及热熔黏合等处理工艺，提高纤维集合体的抗压缩性能，更好地满足浮力材料要求。

吸油拒水材料：由于其疏水亲油性，木棉纤维用作吸油材料和拒水包装纸。

隔热吸声材料：由于其保暖和隔音，木棉纤维可用于房屋隔热层和吸声层的填料。木棉/毛隔热复合材料比毛纤维隔热材料有更好的吸热性和保暖性。

Reinforcement material: Kapok fiber is also used as reinforcement material in composite due to its acid/alkali resistance. The **flexural** modulus and tenacity of composite are significantly improved.

Illumination material: Kapok fiber is applied as **illumination** material due to its high refractive index.

增强材料：由于其耐酸碱，木棉纤维也用作复合材料中的增强材料，改善复合材料的弯曲模量和强度。

增光材料：由于其高折射率，木棉纤维用作增光材料。

Exercises/ 练习

1. What properties of kapok fiber are better than those of cotton fiber, and why?
2. Translate the following Chinese into English.

（1）木棉纤维的细度仅有棉纤维的1/2，中空率却达到86%以上，是一般棉纤维的2~3倍。

（2）木棉织物光洁、生态环保、不透水、不导热，另外，还具有抗菌、防蛀、防霉、轻柔、保暖、吸湿等特点。

（3）木棉纤维可纺性差，一般难以纯纺。采用与棉、黏胶纤维或其他纤维素纤维混纺，可制织光泽和手感良好的服装面料。

（4）木棉纤维的中空率远远高于其他现有纤维，是优良的隔热、隔音、保暖材料。

（5）木棉纤维或织物可用作服装面料、被褥、枕头和靠垫的填充材料、隔热和吸声材料及浮力材料等。

（6）木棉纤维作为一种天然纤维素纤维，具有薄壁、大中空的独特结构，其中空率远远高于其他现有纤维，是优良的隔热、隔音、保暖和浮力材料。

References/ 参考文献

［1］ZHENG Y, WANG J T, ZHU Y F, et al. Research and application of kapok fiber as an absorbing material: A mini review ［J］. Journal of Environmental Sciences, 2015（27）: 21-32.

［2］丁颖, 楼雪君, 胡真迎, 等. 木棉纤维的性能及其应用 ［J］. 产业用纺织品, 2008, 218（11）: 1-3.

［3］孙晓婷. 木棉纤维的性能及其在纺织上的应用 ［J］. 成都纺织高等专科学校学报, 2016, 33（1）: 145-149.

［4］唐爱民, 孙智华, 付欣, 等. 木棉纤维的基本性质与结构研究 ［J］. 中国造纸学报, 2008, 23（3）: 1-5.

Chapter Two   Natural Colored Cotton Fiber/天然彩棉纤维

1   Introduction/前言

Natural colored cotton fiber, which is cultivated by modern biological technology, is a new cellulosic fiber with natural color since the **boll opening**. The **pigment** is **deposite**d in the cell that causes the coloration during the formation and growth of the fiber cell. The more the deposition, the darker the color. The color varies with different genes. Natural colored cotton fiber products do not need to be dyed/**bleach**ed, and they do not need any chemical agents in the proceess, which effectively avoids environmental pollution and harming human body. Like white cotton fiber, natural colored cotton fiber is a **unicellular** seed fiber, which is grown from the seed epidermal cell.

Research and development of natural colored cotton was not started until the early 1960s around the world. In the early 1970s, nine **varieties** of natural colored cotton have been produced and named. Among them, four varieties of natural colored cotton, namely "New colored cotton 1, New colored cotton 2, New colored cotton 3, New colored cotton 4", have been successfully grown in Xinjiang Natural Colored Cotton Research Institution, subsidiary of China Color Cotton Company. A new variety of natural colored cotton was also cultivated in Hunan Cotton Research Institution. At present, the chief natural colored cotton-

天然彩棉纤维是采用现代生物工程技术培育出来的一种在棉花吐絮时纤维就具有天然色彩的新型纤维素纤维。其显色原因主要是在纤维细胞的形成与生长过程中，细胞里沉积了某种色素物质。沉积越多，颜色越深，不同的控制基因形成了多种颜色。天然彩棉纤维产品不需漂染，加工过程中不消耗化学试剂，有效避免对环境的污染和对人体的危害。和白棉纤维一样，天然彩棉纤维是由种子的表皮细胞生长而成的单细胞种子纤维。

20世纪60年代初，全世界开始进行天然彩棉的研发，70年代初，通过审定命名的天然彩棉品种有9个。中国彩棉公司下属的新疆天然彩色棉花研究所成功研发4个品种，分别命名为"新彩棉1、新彩棉2、新彩棉3和新彩棉4"，另外，湖南棉花研究所培育1个品种。目前，中国、美国和巴西等国家进行天然彩棉种植、研发和生产，

growing, cotton-researching and cotton-manufacturing countries of the world are China, America, Brazil and other countries. In China, natural colored cotton has succeefully been grown in Xinjiang, Gansu, Sichuan, Hainan, etc. And the yield, quality and fiber length have all been assessed to come up to the world advanced level. It is grown mostly in Xinjiang. The distribution of natural colored cotton variety is shown in Figure 2-1. It is predicted that the production of natural colored cotton will occupy up to about 30% of the global cotton over next 30 years. About 60%~70% of the world's population will comsume natural colored cotton products in the 21st century. The **yield**s of natural colored cotton and white cotton are shown in Figure 2-2.

Natural colored cotton is usually classified by color and variety. According to the color, it is primarily divided

在我国，新疆、甘肃、四川和海南等地天然彩棉种植已经取得成功，其纤维产量、质量和长度均达到世界先进水平，新疆是我国种植天然彩棉最多的地方，天然彩棉品种分布如图 2-1 所示。预测未来 30 年内，天然彩棉产量将约占全球棉花总产量的 30%。21 世纪，全世界将有 60%~70% 的人口使用天然彩棉制品。早年间，天然彩棉与白棉的产量如图 2-2 所示。

天然彩棉可按颜色和种类进行分类。按颜色分类，天然彩棉

Figure 2-1　Distribution of natural colored cotton varieties
图2-1　天然彩棉品种分布

(a) Natural colored cotton/天然彩棉

Figure 2-2
图2-2

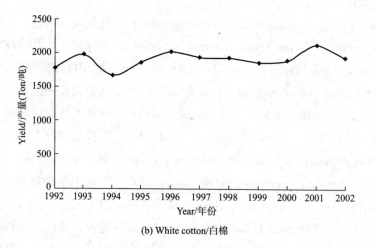

(b) White cotton/白棉

Figure 2-2　Cotton yields
图2-2　棉的产量

into brown cotton and green cotton. According to the variety, it contains land colored cotton, Asia colored cotton, sea-island colored cotton and African colored cotton. The yield of Land colored cotton is the highest, followed by that of Asia colored cotton, and the yields of sea-island colored cotton and African cotton are the lowest.

主要有棕棉和绿棉；按种类分类，天然彩棉有陆地彩棉、亚洲彩棉、海岛彩棉和非洲彩棉，其中陆地彩棉产量最多，亚洲彩棉次之，海岛彩棉和非洲彩棉最少。

2　Structure/ 结构

2.1　Molecular Structure/ 分子结构

As a typical cellulosic fiber, natural colored cotton fiber is chemically composed of cellulose, hemicellulose, wax, **pectin** and so on. The chemical composition of brown, green and white cotton fibers are exhibited in Table 2-1. As is shown, natural colored cotton fiber contains lower cellulose than white cotton fiber. Green cotton fiber contains the highest wax, followed by brown cotton fiber, and white cotton fiber has the lowest wax. The wax on the surface could protect the rest of the fiber against chemical and other degrading agents. The wax content of natural colored cotton fiber appreciably varies with color and variety. Even

作为典型的纤维素纤维，天然彩棉纤维的主要化学成分为纤维素、半纤维素、蜡质和果胶等物质。棕棉、绿棉和白棉纤维的化学组成见表2-1。天然彩棉纤维的纤维素含量低于白棉纤维。绿棉纤维的蜡质含量最高，棕棉纤维次之，白棉纤维最低。表面的蜡质保护纤维各部分免受化学或其他降解剂的影响。不同颜色、不同品种的彩棉纤维中蜡质

with the same color, the wax content increases with the deepening of color **shade**. Brown cotton fiber contains the highest pectin, followed by white cotton fiber, and green cotton fiber has the lowest pectin.

含量有很大的差别，同种颜色彩棉纤维中蜡质含量随着颜色的加深而增大。棕棉纤维的果胶含量最高，白棉纤维次之，绿棉纤维最低。

Table 2–1　Chemical compositions of cotton fibers
表 2–1　棉纤维的化学组成

Chemical composition 化学组成（%）	White cotton 白棉	Brown cotton 棕棉	Green cotton 绿棉
Cellulose 纤维素	89.90~94.93	83.49~86.23	81.09~84.88
Hemicellulose 半纤维素	1.11~1.89	1.35~2.07	1.64~2.78
Wax 蜡质	0.57~0.88	0.67~0.89	3.24~4.69
Pectin 果胶	0.32~0.84	0.46~0.93	0.28~0.81

The **infrared spectra** of brown and white cotton fibers are shown in Figure 2–3. They have the same characteristic absorption band. The absorption peaks at 3400cm^{-1} and 2900cm^{-1} correspond to —OH **stretching vibration absorption band** and —CH stretching vibration absorption band, respectively. The absorption peak at 1630cm^{-1} is

棕棉和白棉纤维的红外吸收光谱如图 2–3 所示。它们具有相同的特征吸收光谱带，3400cm^{-1} 和 2900cm^{-1} 吸收峰分别为纤维素纤维的—OH 伸缩振动吸收带和—CH 伸缩振动吸收带，

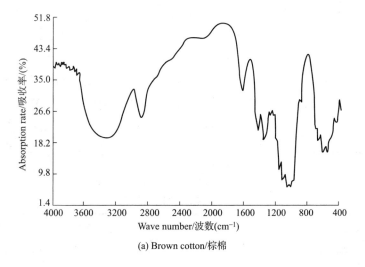

(a) Brown cotton/棕棉

Figure 2–3
图 2–3

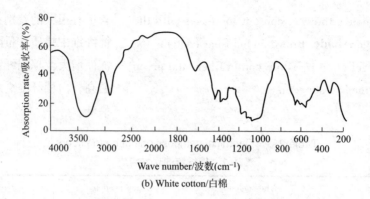

Figure 2-3　Infrared spectra of cotton fibers
图2-3　棉纤维的红外光谱

attributed to the H—O—H absorption band resulting from the moisture absorption of the cellulose fiber.

1630cm^{-1}为纤维素纤维吸收水分而产生的H—O—H吸收带。

2.2　Supramolecular Structure/ 超分子结构

The differences in the color and origin of cotton fibers cause crystallinity and the degree of orientation to vary. The crystallinity of cotton fibers are shown in Table 2-2. The crystallinity of brown, green and white cotton fibers are 75%~81%, 70%~74% and 65%~72%, respectively. Natural colored cotton fiber has higher crystallinity than white cotton fiber. Brown cotton fiber has slightly higher crystallinity than green cotton fiber. In cotton fiber, some polymer chains lie parallel to the fiber axis, but many polymer chains lie at an appreciable angle to the axis. The cellulose polymers **spiral** at 20°~30°. Brown cotton fiber has similar orientation degree to white cotton fiber. And green cotton fiber has a little lower orientation degree than brown and white cotton fibers.

棉纤维的颜色和产地不同，它们的结晶度和取向度有差异。棉纤维的结晶度见表2-2。棕棉、绿棉和白棉纤维的结晶度分别是75%~81%，70%~74%和65%~72%。彩棉纤维的结晶度高于白棉纤维，棕棉纤维的结晶度略高于绿棉纤维。棉纤维中一些大分子链与纤维轴向平行，而大部分大分子链与轴向成一定角度，纤维素大分子取向为20°~30°。棕棉纤维的取向度与白棉纤维相似，绿棉纤维的取向度略小于棕棉和白棉纤维。

Table 2-2　Crystallinity of cotton fibers
表2-2　棉纤维的结晶度

Fiber 纤维	Brown cotton 棕棉	Green cotton 绿棉	White cotton 白棉
Crystallinity 结晶度（%）	75~81	70~74	65~72

2.3 Morphological Structure/ 形态结构

SEM images of the **microstructure** of cotton fibers are shown in Figure 2-4~Figure 2-6. The cross section of cotton fibers is composed of several **concentric cylinder**s, namely **primary wall** (PW), **secondary wall** (SW) and lumen (L) from outside to inside. The SW of natural colored cotton fiber can be divided into the inner layer of SW (ISW) and the outer layer of SW (OSW). No coloration matter has ever been observed in white cotton fiber. The coloration matters of green and brown cotton fibers are deposited in ISW and L, respectively. The microstructure parameters of cotton fibers are shown in Table 2-3. For white cotton fiber, the thickness of PW, SW and L are 0.2 μm, 11 μm and 6.15 μm, respectively. Natural colored cotton fiber has thinner SW than white cotton fiber. Green cotton fiber has thicker SW than brown cotton fiber. Brown cotton fiber has considerably thicker L than white and green cotton fibers. Therefore, white cotton fiber has the largest diameter, followed by brown cotton fiber, and green cotton fiber has the smallest diameter.

棉纤维微观结构的电子扫描显微镜照片如图2-4~图2-6所示，棉纤维的横截面是由许多同心圆柱组成，由外至内依次为初生层、次生层和中腔。天然彩棉纤维的次生层分为次生层内层和次生层外层。白棉纤维无色素，而绿棉和棕棉纤维的色素分别沉积在次生层内层和中腔内。棉纤维的微观结构参数见表2-3，白棉纤维的初生层、次生层和中腔的厚度分别为0.2 μm、11 μm和6.15 μm，彩棉纤维的次生层厚度远小于白棉纤维，绿棉纤维的次生层厚度大于棕棉纤维，棕棉纤维的中腔厚度远大于白棉和绿棉纤维，因此，白棉纤维的直径最大，棕棉纤维次之，绿棉纤维最小。

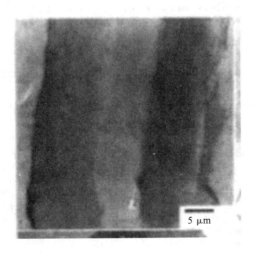

Figure 2-4 Microstructure of white cotton fiber
图2-4 白棉纤维的微观结构

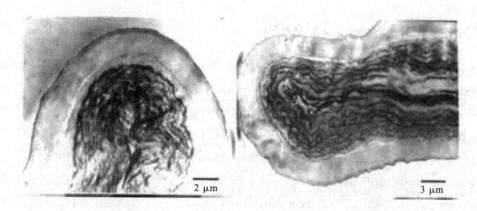

Figure 2-5　**Horizontal** and **vertical** cross sections of green cotton fiber
图2-5　绿棉纤维的横和纵截面

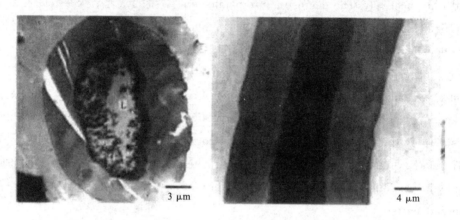

Figure 2-6　Horizontal and vertical cross sections of brown cotton fiber
图2-6　棕棉纤维的横和纵截面

Table 2-3　Microstructure parameters of cotton fibers
表 2-3　棉纤维的微观结构参数

Fiber 纤维	Coloration matter 色素	Thickness 厚度（μm）			Diameter 直径（μm）	
		PW 初生壁	SW 次生胞壁		L 中腔	Fiber 纤维
			OSW 次生壁外层	ISW 次生壁内层		
White cotton 白棉	No 无	0.2	11		6.15	28.55
Green cotton 绿棉	Green in ISW 次生壁内层含绿色素	0.1	2.1	3.6	1.2	12.8
Brown cotton 棕棉	Brown in L 中腔内含棕色素	0.1	3.7		10.3	17.9

As is shown in Figure 2-7, natural colored cotton fiber has similar cross sectional and longitudinal morphology structures to white cotton fiber. The cross section of natural colored cotton fiber is kidney-shaped with the lumen larger than that of white cotton fiber. Natural colored cotton fiber appears as a flat ribbon with irregular natural convolution in the direction along the fiber length. The number of convolution in natural colored cotton fiber is less than that in white cotton fiber. The number of natural convolution in brown cotton fiber is shown in Table 2-4.

如图 2-7 所示，天然彩棉纤维的横截面和纵向形态与白棉纤维相似。横截面呈腰圆形，中腔大于白棉纤维。纵向是不规则转曲的扁平带状，天然彩棉纤维的纵向转曲数比白棉纤维少。棕棉纤维的天然转曲数见表 2-4。

 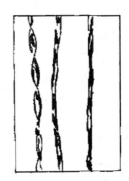

Figure 2-7 Cross section and longitudinal morphology of natural colored cotton fiber
图2-7 天然彩棉纤维的横截面和纵向形态

Table 2-4 Number of natural convolution in brown cotton fiber
表 2-4 棕棉纤维的天然转曲数

Part 部位	Root 根部	Center 中部	Tip 梢部
Forward 正向	5~10	7~11	4~9
Reverse 反向	1~2	2~3	1~2

Mature natural colored cotton fiber appears as a full ribbon with more convolutions and small lumen, whereas immature natural colored cotton fiber appears as a **tabular** ribbon with less convolutions and large lumen. The coarser natural colored cotton fiber is more porous than the finer one.

成熟度好的天然彩棉纤维纵向呈转曲的带状，转曲数较多，中腔较小；成熟度较差的呈扁平带状，转曲数较少，中腔较大。较粗的彩棉纤维气孔比较细的纤维更多。

3 Property/ 性能

The physical properties of cotton fibers are summarized in Table 2-5 and Table 2-6.

棉纤维的物理性能见表 2-5 和表 2-6。

Table 2-5 Physical properties of cotton fibers
表 2-5 棉纤维的物理性能

Variety 种类	Green cotton 绿棉	Brown cotton 棕棉	White cotton 白棉
Upper half mean length 上半部平均长度（mm）	21~25	20~23	33~35
2.5% Span length 2.5%跨距长度（mm）	21~25	20~23	28~31
Breaking tenacity 断裂强度（cN/tex）	1.6~1.7	1.4~1.6	1.9~2.3
Moisture regain 回潮率（%）	7.98	8.42	8.40
Micronaire 马克隆值	3.0~6.0	3.0~6.0	3.7~5.0
Uniformity 整齐度（%）	45~47	44~47	49~52
Short fiber ratio 短绒率（%）	15~20	15~30	≤12
Neps 棉结（g）	100~150	120~200	80~200
Lint ratio 衣分率（%）	20	28~30	39~41

Table 2-6 Length and tenacity of natural colored cotton fiber grown in the United States
表 2-6 美国天然彩棉纤维的长度和强度

Species 品种	Length 长度（mm）	Tenacity 强度（cN/tex）
Wild Green 浅绿	19.6	1.27
Old Green 原绿	23.9	1.66
Palo Verde Green 绿色	23.9	1.85

续表

Species 品种	Length 长度（mm）	Tenacity 强度（cN/tex）
New Green 新绿	29.0	2.33
Wild Brown 浅棕	17.5	1.37
Coyote Brown 狼棕	23.9	1.96
Buffalo Brown 水牛棕	27.7	2.45
New Brown/新棕	27.4	2.49

Length and fineness: The length of natural colored cotton fiber is 2/3~4/5 that of white cotton fiber, which ranges from 28 mm to 31 mm. Green and brown cotton fibers range from 21 mm to 25 mm and 20 mm to 23 mm in length, respectively. Green cotton fiber is slightly longer than brown cotton fiber. Natural colored cotton fiber has slightly higher linear density (1.22~2.45 dtex) than white cotton fiber (1.22~2.04 dtex).

Tensile property: The breaking tenacity of green and brown cotton fibers is 1.6~1.7 cN/dtex and 1.4~1.6 cN/dtex, respectively, which is lower than that of white cotton fiber (1.9~2.3 cN/dtex). The crystallinity and noncellulosic substance content influence the fiber tensile property. The higher the crystallinity is, the higher the breaking tenacity is. Though the crystallinity in natural colored cotton fiber is higher than it is in white cotton fiber, natural colored cotton fiber has lower tenacity than white cotton fiber. This might be attributed to the high content of noncellulosic substances (including pigment) in natural colored cotton fiber. The **resilience** of natural colored cotton fiber is low. Natural colored cotton fabrics therefore are susceptible to **wrinkle** during using and laundering, and have to be **iron**ed unless they have a **durable-press finishing**.

长度和细度：白棉纤维长度为28~31 mm，天然彩棉纤维长度为白棉的2/3~4/5。绿棉和棕棉纤维的长度分别为21~25 mm和20~23 mm，绿棉纤维比棕棉纤维略长。天然彩棉纤维的线密度为1.22~2.45 dtex，略高于白棉纤维的线密度（1.22~2.04 dtex）。

拉伸性：绿棉和棕棉纤维的断裂强度分别为1.6~1.7 cN/dtex和1.4~1.6 cN/dtex，均低于白棉纤维的断裂强度（1.9~2.3 cN/dtex）。纤维的结晶度和非纤维素物质含量影响其强伸性，结晶度高的纤维，纤维断裂强度越高。虽然天然彩棉纤维的结晶度高于白棉纤维，但是断裂强度低于白棉纤维，可能与彩棉纤维中较高杂质（包括色素）含量有关。彩棉纤维的弹性回复率很低，彩棉织物如未经耐用性后整理，在使用和洗涤过程中容易起皱，须熨烫处理。

Spinnability: The chemical composition that primarily affects the spinnability of cotton fiber is cellulose. Natural colored cotton fiber contains considerably lower cellulose and remarkably higher hemicellulose content than white cotton fiber, leading to the poor spinnability of natural colored cotton fiber. The spinnability of brown cotton fiber is relatively better than that of green cotton fiber due to the slightly higher cellulose content in brown cotton fiber than in green cotton fiber. The pectin content in natural colored cotton fiber only accounts for 35%~45% of that in white cotton fiber, which lowers the **cohesion** between cell walls, deteriorates the spinnability as well as increases the **breakage** and **fuzziness** in the spinning process.

The spinning technology and yarn quality of natural colored cotton fiber are appreciably affected by short fiber length, low breaking tenacity and uniformity as well as high short fiber content. Natural colored cotton fiber has higher **impurity** content (up to 9.8% at the highest) than white cotton fiber. In the ginning process, **sterile seed**s are crushed down into **chip**s, which is hard to be removed in spinning and finally become harmful **defect**s on the fabric surface.

Moisture absorption: The wax and pectin have great influence on the moisture absorption of cotton fiber. The higher the wax content is, the worse the moisture absorption is due to the hydrophobicity of wax. To the contrary, the higher the pectin content is, the higher the moisture absorption is due to the **hydrophilicity** of pectin. Green cotton fiber contains exceptionally higher wax than brown and white cotton fibers. And brown cotton fiber contains higher pectin than white cotton fiber. The moisture regain of brown cotton fiber is the highest, followed by that of white cotton fiber, and the moisture regain of green cotton fiber is the lowest.

Dimensional stability: Natural cotton fiber has poor dimensional stability. It is more inclined to **shrink**

可纺性：影响棉纤维可纺性的主要化学成分为纤维素。天然彩棉纤维的纤维素含量远远小于白棉，而半纤维素含量较多，造成彩棉纤维可纺性差。棕棉纤维素含量略高于绿棉，因此，棕棉可纺性相对好些。彩棉纤维果胶含量仅为白棉的35%~45%，彩棉胞壁之间的抱合力较低，因此，可纺性差，在纺纱过程中，断头率高，易起毛。

彩棉纤维长度较短，断裂强度和整齐度较低，短绒含量较高，严重影响纺纱工艺和成纱质量。彩棉纤维的杂质含量高于白棉纤维，最高时可达9.8%，不孕籽在轧花过程中，被机器轧碎后成为带纤维籽屑，纺纱中极难清除，成为织物表面的有害疵点。

吸湿性：蜡质和果胶对棉纤维吸湿性的影响很大。蜡质含量越高，由于其疏水性使得棉纤维吸湿性越差。相反，果胶含量越高，由于其亲水性使得棉纤维吸湿性越好。绿棉纤维的蜡质含量比棕棉和白棉纤维高得多。棕棉纤维的果胶含量高于白棉纤维。棕棉纤维回潮率最高，白棉纤维次之，绿棉纤维最低。

尺寸稳定性：彩棉纤维次生壁薄，中腔大，易收缩和变形，

and **deform** due to its thin secondary wall and large lumen. After hot water treatment for 1 h at 50~60 ℃, the **shrinkage ratio** of cross sectional area of natural colored cotton fiber reaches more than 16%, and after concentrated alkali treatment, it is as high as 30%. The lumen in cotton fiber will disappear after being mercerized, resulting in better dimensional stability.

Air permeability and warmth retention: Natural colored cotton fiber has excellent air permeability and warmth retention owing to its lumen structure, through which the moisture can be freely expelled and the static air can be held.

Antistatic property: On account of the high moisture regain of natural colored cotton fiber, it is resistant to generate **static electricity**. It has good antistatic property.

Ultraviolet (UV) proof: The pigment in natural colored cotton fiber has a strong absorption of ultraviolet light and a certain shielding effect on UVA and UVB. Natural colored cotton fiber has a remarkably lower UV **transmittance** than white cotton fiber.

Dyeability: Because of its higher crystallinity, natural colored cotton fiber has poorer dyeability than white cotton fiber. In the dyeing and finishing processes, chemical agents generally are used in white cotton fiber to remove the wax. Then the white cotton fiber could be easily dyed. No chemical agents are required to be applied in natural colored cotton fiber during processing. Hence the characteristics of natural fiber are well preserved. However, natural colored cotton fiber has lower color brightness than the dyed white cotton fiber.

Color fastness: After being washed, the color of natural colored cotton fiber deepens instead of fading out. The color becomes darker with the rise of water temperature and the increase of washing times. Natural colored cotton fiber has poor color fastness to sunlight. Prolonged exposure to sunlight causes green cotton fiber to yellow-green and

尺寸稳定性差。经50~60℃热水处理1 h，彩棉纤维横截面面积收缩率可达到16%以上。经过浓碱处理，可达到30%以上。棉纤维经丝光处理后，胞腔基本消失，从而获得较好的尺寸稳定性。

透气保暖性：彩棉纤维有较大的中腔，水蒸气可自由排出，同时能容纳较多的静止空气，达到极好的透气和保暖效果。

抗静电性能：彩棉纤维的回潮率较高，不易起静电，抗静电性好。

防紫外线性：彩棉纤维的色素对紫外线有强烈的吸收性，对UVA和UVB均有一定的屏蔽作用，紫外线透过率显著低于白棉纤维。

染色性：彩棉纤维结晶度较高，染色性比白棉纤维差。白棉纤维在印染和后整理过程中，使用了一些化学试剂，蜡质被去除，容易着色。彩棉在加工过程中未使用化学试剂处理，仍旧保留天然纤维的特点，但鲜亮度不如染色的白棉纤维。

色牢度：天然彩棉纤维洗涤后不会褪色反而加深，随水洗温度及水洗次数的增加，颜色逐渐加深。但是耐日晒牢度差，在长时间自然光照情况下，绿棉纤维的颜色会随时间的推移，渐向黄

gradually yellow-brown.

Luster: The luster of natural colored cotton fiber is weak. The convoluted fiber surface reflects light in a scattered pattern. The low luster of natural colored cotton fiber and the fuzziness of yarns spun from this fiber result in low-luster fabrics.

Acid/alkali resistance: Strong acid degrades natural colored cotton fiber. Hot, dilute acid causes it to disintegrate. Cold, dilute acid leads to gradual degradation, which is slow and may not be immediately evident.

Antibacterial activity: Natural colored cotton fiber is susceptible to be damaged by fungus like **mildew**, and by **bacterium**.

绿色、黄棕色转变。

光泽：天然彩棉纤维的光泽度低，转曲的纤维表面分散地反射光线。彩棉纤维的低光泽度和纱线的起毛性决定了其织物具有较低的光泽。

耐酸碱性：天然彩棉纤维遇强酸降解，遇热稀酸降解，遇冷稀酸缓慢降解，不易察觉。

抗菌性：天然彩棉纤维容易被真菌（如霉菌）及细菌破坏。

4 Scouring Process/ 煮练工艺

The moisture absorption of cotton fiber is greatly influenced by wax. The wax and cellulose are boned together by pectin which acts as an **adhesive**. Alkali or **pectinase** treatment is usually applied to remove pectin so that wax is seperated from cellulose. The obtained cotton is so called **degreasing cotton** (**absorbent cotton**).

棉纤维中的蜡质大大影响其吸湿性。蜡质和纤维素被作为黏合剂的果胶结合在一起，通过碱或果胶酶处理棉纤维去除果胶，使蜡质和纤维素分离。去除脂肪的棉纤维称为脱脂棉纤维。

4.1 Evaluation of Pectin Content/ 果胶含量测定

Extraction method is used to positively evaluate the pectin content with **ammonium oxalate** as the **extractant**. The pectin content in cotton fiber is calculated according to Equation (2-1). Results are listed in Table 2-7. As is shown, natural colored cotton fiber has almost the same pectin content like white cotton fiber.

采用草酸铵作为萃取剂，用萃取法测定棉纤维中的果胶含量。按公式（2-1）可计算得到棉纤维中果胶含量。棉纤维的果胶含量见表2-7。天然彩棉和白棉纤维的果胶含量基本一致。

$$P = \frac{(W_o - W_d)}{W_o} \times 100\%$$

（2-1）

Where, P is Percentage of pectin; W_o (g) is weight before extraction; W_d (g) is weight after extraction。

式中：P 为果胶含量（%）；W_o 为萃取前质量（g）；W_d 为萃

Table 2-7 Pectin contents of cotton fibers
表 2-7 棉纤维的果胶含量

Temperature 温度（℃）	Pectin content 果胶含量（%）		
	White cotton 白棉纤维	Brown cotton 棕棉纤维	Green cotton 绿棉纤维
70	1.09	1.49	1.52
90	2.84	2.33	2.37

取后质量（g）。

4.2 Alkali Scouring/ 碱煮练

The color values and wicking height of the natural colored cotton fabric scoured by alkali are shown in Table 2-8. Initial color values for L, a and b are 40.36, 23.83 and 10.50, respectively. And the wicking height is zero. After alkali scouring, the color values and wicking height are as follows.

(1) The color is obviously darker in the natural colored cotton fabric scoured by alkali than in the control.

(2) With the increasing soda ash concentration, scouring temperature and scouring time, the moisture absorption of the natural colored cotton fabric scoured by alkali increases and its color becomes darker.

(3) The optimal processing parameters for the natural colored cotton fabric scoured by alkali are: soda ash

碱煮练的天然彩棉织物的颜色值和芯吸高度见表 2-8。原样颜色值 L、a 和 b 分别为 40.36、23.83 和 10.50，芯吸高度为零。碱煮练的天然彩棉织物的颜色值和芯吸高度如下。

（1）碱煮练的天然彩棉织物的颜色比原样深。

（2）随着纯碱浓度、煮练温度和煮练时间的增加，天然彩棉织物的吸湿性提高，颜色变深。

（3）天然彩棉织物碱煮练的最佳工艺参数为：纯碱浓度为

Table 2-8 The color values and wicking height of the natural colored cotton fabric scoured by alkali
表 2-8 碱煮练的天然彩棉织物的颜色值和芯吸高度

Parameter 参数		Soda ash concentration 纯碱浓度（g/L）			Scouring temperature 煮练温度（℃）		Scouring time 煮练时间（min）	
		1	3	5	70	100	30	60
Color 颜色	L	35.31	34.02	33.62	34.02	33.30	34.02	33.86
	a	24.52	25.05	25.26	25.05	26.18	25.05	25.17
	b	10.18	9.74	7.46	9.74	4.77	9.74	8.20
Wicking height 芯吸高度（cm）		6.3	11.7	13.7	11.7	14.3	11.7	12.9

concentration: 3 g/L, scouring temperature: 70 ℃ and scouring time: 30 min.

3 g/L，煮练温度为 70 ℃，煮练时间为 30 min。

4.3 Pectase Scouring/ 果胶酶煮练

Different pectases are considered to be performed in solutions with different pH values. For example, the pH value of the solution should be above 7 when the special pectase for cotton fiber is chosen. The pectase named Bioprop supplied by Denmark Novo Nordisk China Co. Ltd. requires the pH value of the solution ranging from 8.0 to 10.0.

不同果胶酶在不同的 pH 溶液中使用。例如，在选择棉花煮练专用果胶酶时，溶液 pH 应大于 7。丹麦 Novo Nordisk 中国有限公司提供的果胶 Bioprop 要求溶液 pH 范围为 8.0~10.0。

The color values and wicking height of natural colored cotton fabrics scoured by pectase are shown in Table 2-9. Initial color values for L, a and b are 40.36, 23.83 and 10.50, respectively. And the wicking height is zero. After pectase scouring, the color and wicking height are as followings.

果胶酶煮练的天然彩棉织物的颜色值和芯级高度见表 2-9。原样颜色值 L、a 和 b 分别为 40.36、23.83 和 10.50，芯吸高度为零。果胶酶煮练的天然彩棉织物的颜色值和芯吸高度如下。

(1) The color is slightly darker in natural colored cotton fabrics scoured by pectase than in the control.

（1）果胶酶煮练的天然彩棉织物的颜色比原样略深。

(2) With the increasing pectase concentration, scouring temperature and scouring time, the moisture absorption of natural colored cotton fabric scoured by pectase improves and its color becomes darker. The pectase concentration turns out more crucial to the fabric than scouring time. The

（2）随着果胶酶浓度的提高和煮练时间的增加，果胶酶煮练的天然彩棉织物吸湿性增加，颜色变深。果胶酶浓度对天然彩棉织物的影响大于煮练时间。去除

Table 2-9 The color values and wicking height of natural colored cotton fabrics scoured by pectase
表 2-9 果胶酶煮练的天然彩棉织物的颜色值和芯级高度

Parameter 参数		Pectase concentration 果胶酶浓度（g/L）			Scouring temperature 煮练温度（℃）			Scouring time 煮练时间（min）		
		0.01	0.05	0.1	Room temperature 室温	55		10	30	60
Color 颜色	L	38.57	37.98	37.71	39.02	37.98		40.02	37.98	37.81
	a	23.58	23.96	24.44	23.89	23.96		23.97	23.96	24.08
	b	9.24	9.02	8.88	9.37	9.02		9.96	9.02	8.91
Wicking height 芯吸高度（cm）		0	8.7	9.2	0	8		0	8.7	8.9

rate at which the pectin is removed from natural colored cotton fabric is mostly decided by pectase concentration. In addition, the quantity of the removed pectin is in proportional to pectase concentration.

(3) The optimal processing paramters for natural colored cotton fabrics scoured by pectase are: pectase concentration: 0.05% (in weight), scouring temperature: 55 ℃ and scouring time: 30 min.

4.4 Comparisons of Alkali and Pectase Scouring/ 碱煮练和果胶酶煮练对比

Comparisons of natural colored cotton fabrics scoured by alkali and pectase are shown in Table 2-10. Alkali scouring has better scouring effect than pectase scouring. The color is darker in the natural clored cotton fabric scoured by alkali than in that scoured by pectase. The moisture absorption of the natural clored cotton fabric scoured by alkali is better than that of that scourde by

Table 2-10　Comparisons of alkali and pectase scouring effects

表 2-10　天然彩棉织物的碱煮练和果胶酶煮练效果比较

Condition 条件			100% Grown cotton 棕棉	25% Brown cotton 棕棉	100% Green cotton 绿棉	33% Green cotton 绿棉
Control 原样	Color 颜色	L	40.36	57.97	63.00	67.09
		a	23.83	19.15	3.94	5.28
		b	10.50	13.00	12.50	12.25
	Wicking height 芯吸高度（cm）		0	0	0	0
Alkali scouring 碱煮练	Color 颜色	L	34.02	54.05	54.51	61.23
		a	25.05	16.03	0.62	−0.66
		b	9.74	9.82	8.87	9.84
	Wicking height 芯吸高度（cm）		11.7	11.0	12.2	11.6
Pectase scouring 果胶酶煮练	Color 颜色	L	37.98	55.86	60.39	64.72
		a	23.96	18.21	−3.05	−1.37
		b	9.02	10.95	11.30	10.06
	Wicking height 芯吸高度（cm）		8.7	8.1	8.8	8.3

pectase. However, the damage of natural colored cotton fabric caused by alkali scouring is greater than that by pectase scouring. Therefore, the natural clored cotton fabric scoured by alkali has lower tenacity than that scoured by pectase.

织物的损伤大于果胶酶煮练，因此，碱煮练的天然彩棉织物强度比果胶酶煮练的低。

4.5 Color Fastness of Natural Colored Cotton Fabrics After Alkali and Pectase Scouring/ 碱煮练和果胶酶煮练的天然彩棉织物的色牢度

According to GB 414—78, GB 415—78 and GB 416—78, the washing color fastness, rubbing color fastness and perspiration color fastness of natural colored cotton fabrics scoured by alkali and pectase are tested, respectively. Color fastness of natural colored cotton fabrics scoured by alkali and pectase are shown in Table 2–11.

根据 GB 414—78、GB 415—78 和 GB 416—78 分别测试碱煮练和果胶酶煮练的天然彩棉织物的水洗色牢度、摩擦色牢度和汗渍色牢度。碱煮练和果胶酶煮练的天然彩棉织物的色牢度见表 2–11。

Table 2–11　Color fastness of the natural colored cotton fabrics scoured by alkali and pectase
表 2–11　碱煮练和果胶酶煮练的天然彩棉织物的色牢度

Condition 条件		Washing 水洗		Rubbing 摩擦		Perspiration (acid) 耐汗（酸）		Perspiration (alkali) 耐汗（碱）	
		Fading 褪色	Staining 沾色	Dry 干	Wet 湿	Fading 褪色	Staining 沾色	Fading 褪色	Staining 沾色
Brown cotton 棕棉	Alkali 碱	4	4–5	3–4	2–3	4	4–5	4	4–5
	Pectase 果胶酶	3–4	4–5	3–4	2	4	4–5	3–4	4–5
Green cotton 绿棉	Alkali 碱	4	4–5	3–4	2–3	4	4–5	4	4–5
	Pectase 果胶酶	3–4	4–5	3–4	2	4	4–5	3–4	4–5

(1) For washing color fastness, the natural colored cotton fabric scoured by pectase is graded lower than that scoured by alkali.

(2) For rubbing color fastness, the scoured natural colored cotton fabric has a higher grade when dry than wet. The natural colored cotton fabric scoured by pectase is

（1）对于水洗色牢度，果胶酶煮练的天然彩棉织物的水洗色牢度等级低于碱煮练的。

（2）对于摩擦色牢度，煮练的天然彩棉织物的干摩擦色牢度等级高于湿摩擦色牢度等级，果

graded lower than that scoured by alkali when wet. However, they have the same dry rubbing color fastness.

(3) For perspiration color fastness, natural colored cotton fabric scoured by pectase is graded lower than that scoured by alkali under alkaline condition. However, they have the same dry perspiration color fastness under acidic condition.

(4) The staining of natural colored cotton fabrics scoured by pectase is not very obvious. The staining is usually graded 4–5.

胶酶煮练的天然彩棉织物的湿摩擦色牢度低于碱煮练的，但他们的干摩擦色牢度等级相同。

（3）对于汗渍色牢度，在碱性条件下，果胶酶煮练的天然彩棉织物的汗渍色牢度等级低于碱煮练的，但在酸性条件下，果胶酶煮练的天然彩棉织物的汗渍色牢度等级与碱煮练的相同。

（4）果胶酶煮练的天然彩棉织物沾色都很轻，等级为4-5级。

5 Fiber Identification/ 鉴别

5.1 Washing Method/ 洗涤法

Immerse the sample in water, and then wash, dry and observe it. If the sample color deepens and becomes more vivid, it is the natural colored cotton fiber. The more the natural colored cotton fiber is washed, the more vivid the color will become. If the sample color **fade**s, then it is the dyed white cotton fiber.

样品浸泡水中，然后清洗、干燥和观察。如果样品颜色变深、鲜艳，则是彩棉纤维。因为天然彩棉纤维洗涤次数越多，纤维表面颜色则越鲜艳。如果颜色变浅，则是染色的白棉纤维。

5.2 Sectioning Method/ 切片法

Make the cross sectional chip of the sample and observe it under the optical microscope. If the color of cross sectional chip gradually fades from the center to the edge, the sample is the natural colored cotton fiber. Conversely, it is the dyed white cotton fiber.

制作样品的横截面切片，并用光学显微镜观察。如果切片的颜色从中心到边缘逐渐变浅，则样品是天然彩棉纤维。如果切片的颜色从中心到边缘逐渐变深，则样品是染色的白棉纤维。

6 Application/ 应用

Natural colored cotton fiber provides unique properties that lead to acceptable fabric performance in apparel and household textiles.

Apparel textile: Since natural colored cotton fiber doesn't need to be dyed or bleached and exhibits excellent behaviors of moisture absorption, air permeability, heat perservation and so on, it is exclusively used as apparel textiles, such as underwear that contacts directly with the skin, infant and woman textiles, and other **casual** clothes like sports clothes, T-shirt, knitted sweater, etc. Natural colored cotton fiber absorbs water vapor emitted by the body, so the skin does not become wet.

Household textile: With advanced weaving technology, natural colored cotton fiber could be applied in household textiles such as bedsheet, quilt cover, pillow case, cushion cover, bath towel, towel, etc.

由于其独特的服用性,天然彩棉纤维主要用于服装和家用纺织品。

服装用纺织品:天然彩棉纤维不需要漂白染色,并具有吸湿、透气和保暖等优点,适合用于服装用纺织品,例如,直接与皮肤接触的内衣、婴幼儿和妇女用纺织品以及运动服、T恤和针织衫等休闲服。天然彩棉纤维可以吸收人体排出的水蒸气,使得皮肤不变湿。

家用纺织品:通过先进的织造工艺,天然彩棉纤维可以用于家用纺织品,例如,床单、被套、枕套和靠垫套等床品及浴巾和毛巾等。

7 Problem/ 问题

(1) Difficult to be grown, low lint ratio and low yield.

Compared with white cotton, natural colored cotton is difficult to be grown and the lint ratio is low, leading to the low yield of natural colored cotton. Besides, it is necessary to take strict measures to prevent natural colored cotton from **hybridization** with white cotton. Otherwise, natural colored cotton fiber will lose its use value. The quality of white cotton will be reduced, as well, due to the pollution of natul colored cotton.

(1)种植难,衣分率低,产量低。

与白棉相比,天然彩棉种植难、衣分率低,造成天然彩棉的产量很低。种植天然彩棉需要采取严格的措施,防止与白棉混杂,否则,天然彩棉将失去使用价值,受天然彩棉的污染,白棉品质也降低。

(2) Poor quality.

Due to its short length, low fineness and tenacity, poor uniformity as well as high content of short fiber, the spinnability of natural colored cotton fiber is inferior to that of white cotton fiber.

(3) Unstable transmissibility.

Natural colored cotton is genetically grown in different color shades due to its unstable **transmissibility**, which affects the color uniformity of products made from it. The color and the quality of natural colored cotton fiber have an inverse relationship. The lighter the color, the better the quality, which is closer to that of white cotton fiber. The darker the color is, the lower the yield and quality are.

(4) Unstable process and use.

The coloration matter in natural colored cotton fiber is unstable. Natural colored cotton fiber therefore easily **stain**s, **discolor**s and fades in case of chemical agents such as acid, alkali, **oxidant**, **reducer**, biological **enzyme**, **penetrant**, **detergent**, and in case of treatments such as soaping, **soften**ing, non-ironing, high temperature and water soaking. The color varies with the changing treatment parameters such as treatment time and temperature, and solution concentration.

（2）品质差。

与白棉纤维相比，天然彩棉纤维品质差，比如，长度短、细度细、强度低、整齐度差、短绒含量高，可纺性不如白棉纤维。

（3）遗传性不稳定。

天然彩棉的遗传性不稳定，使得天然彩棉纤维的颜色深浅不一，影响天然彩棉制品色彩的均匀性。天然彩棉纤维色彩与品质呈反比，颜色越浅，品质越好，越接近于白棉纤维；颜色越深，产量和品质越低。

（4）加工和使用不稳定。

天然彩棉纤维中色素不稳定，遇酸、碱、氧化剂、还原剂、生物酶、渗透剂及洗涤剂等化学试剂以及经皂煮、柔软、免烫、高温和水浸等处理，天然彩棉纤维容易沾色、变色和褪色，颜色随处理时间、温度和溶液浓度等处理参数的变化而变化。

Exercises/ 练习

1. What is natural colored cotton fiber?
2. Please compare diameters of white, brown and green cotton fibers.
3. Please compare the moisture absorption of white, green and brown cottons and give an explanation at the structure level.
4. How to identify the authenticity of natural colored cotton fiber and dyed white cotton fiber?

References/ 参考文献

[1] 张镁，胡伯陶. 彩棉纤维的结构特点与加工性能[J]. 纺织学报，2004, 10（25）: 8-9.

[2] 谭燕玲，周献珠. 天然彩棉研究现状及其发展趋势[J]. 纺织科技进展，2015（2）: 1-4.

[3] 王建坤，潘霞. 天然彩色棉及其工艺性能综述[J]. 天津工业大学学报，2004, 23（4）: 32-36.

[4] 张春娥. 浅析彩棉在舞台服装中的运用[J]. 山东纺织经济，2009, 5（153）: 99-101.

[5] XU W. Scouring process of natural colored cotton products[J]. Journal of Donghua University: English Edition[J]. 2002, 19（1）: 60-64.

[6] 陈莉，黄故，许鹏俊. 碱处理和果胶酶处理彩棉纤维吸湿性能的比较[J]. 天津工业大学学报，2005, 24（3）: 28-30.

Chapter Three Bamboo Fiber/ 竹纤维

1 Introduction/ 前言

Bamboo fiber is a cellulosic fiber obtained from natural bamboo. As the fifth natural fiber after cotton, flax, wool and silk, it features excellent air permeability, instant water absorption, **abrasion resistance** and dyeability, etc. In addition, bamboo fiber exhibits unique functions such as **antibiosis**, **bacteriostasis**, **anti-mite**, **deodorization** and **anti-UV**. As a new environmentally friendly fiber, the demand for its fibre products is increasing year by year.

竹纤维是从天然生长的竹子中提取出来的一种纤维素纤维，是继棉、麻、毛、丝之后的第五大天然纤维。竹纤维具有良好的透气、瞬间吸水、耐磨和染色等特性，同时，又具有天然抗菌、抑菌、除螨、防臭和防紫外线等功能。作为一种新的环保型绿色纤维，其纤维制品需求量逐年上升。

1.1 Bamboo Application/ 竹子用途

Bamboo belongs to **perennial gramineous** plant with a wide variety. Due to its strong **renewability**, short growth period and high output, it is believed that bamboo would be sustainably used with a bright prospect.

竹子是多年生禾本科植物，种类繁多。竹类资源再生能力强，生长周期短，产量高，可持续利用，具有广阔的开发利用前景。

The annual output of bamboo in China ranks first in the world. Bamboo products have already been applied to every aspect in daily life. Abundant bamboo resource provides excellent resource condition for processing and using bamboo, and industrial development in China. It can be used as braided material, such as reinforcement materials including bamboo **plywood**, bamboo fiber board, and so on. It can also be made into decoration materials in

我国竹子年产量列世界首位。竹制品已广泛应用于日常生活的各个领域，丰富的竹类资源为我国竹类加工利用及产业化开发提供了良好的资源条件。竹子可用作编织材料，包括竹编胶合板、竹纤维板等增强材料；建筑、装饰、包装以及交通等装修

construction, decoration, packing and transportation. All kinds of middle and high-grade paper **pulp** can generally be produced from bamboo fiber because it is slender and pliable. The pulping ratio of bamboo is higher than any other plants. Bamboo is extensively used in landscape garden and environmental protection industry due to its elegant shape. Besides, it is of **officinal** value mentioned in ancient Chinese medical books. In recent years, bamboo has extensively been applied in textiles. The development of bamboo fiber and its textiles will open up a new way for bamboo product processing.

材料。竹纤维细长、韧性好，制浆率高于其他木草原料，可制各种中高档纸浆；竹子形态多姿，被广泛应用于风景园林和环保行业；我国历代医学书籍均提到竹子的药用价值。近年来，竹子广泛应用于纺织领域，竹纤维及其纺织品的开发将为竹产品加工开辟一条新途径。

1.2 Bamboo Structure/ 竹子结构

As shown in Figure 3-1, bamboo is **cylindrical**-shaped and consists of three parts including **bamboo stalk**, bamboo joint and bamboo wall. Bamboo stalk has different number of **node**s and **internode** lengths, which are directly related to the origin and variety of the plant. Bamboo joint is composed of stalk ring, **sheath** (around bamboo joints), bud and **nodal diaphragm**). Bamboo wall includes bamboo green, bamboo meat and bamboo yellow.

竹子的基本结构如图3-1所示，竹子呈圆柱形外观，由竹茎、竹节和竹壁组成。不同产地和品种的竹茎有不同节点数和节间长度。竹节由茎环、箨环（叶鞘）和节隔（膜片）组成。竹壁由竹青、竹肉和竹黄组成。

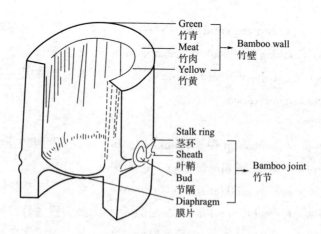

Figure 3-1 Basic structure of bamboo
图3-1 竹子的基本结构

1.3 Fiber Distribution in Bamboo/ 竹子中纤维分布

Figure 3-2 shows the bamboo fiber contents in vertical and horizontal directions of bamboo. In a longitudinal view, the fiber content increases gradually from root to tip. In a horizontal view, it rapidly decreases from bamboo green to bamboo yellow.

竹子纵横向的竹纤维含量如图 3-2 所示。纵向看，从竹子的基部到梢部，竹纤维含量逐渐增加；横向看，从竹青到竹黄，竹纤维含量迅速减少。

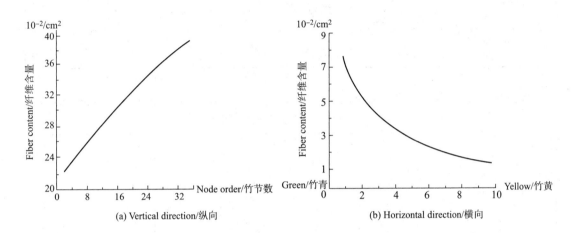

Figure 3-2 Fiber content in bamboo
图3-2 竹子中的纤维含量

1.4 Fiber Length and Fineness in Bamboo/ 竹子中纤维长度和细度

Variations in length and fineness of bamboo fiber with the height of bamboo stalk and with the thickness of bamboo wall are shown in Table 3-1 and Table 3-2, respectively. With the changes in the height of bamboo stalk, the fiber in middle is the longest, followed by that in root and tip. And the fiber in middle is the coarsest, followed by that in tip and root. With the changes in the thickness of bamboo wall, the middle has the longest fiber, followed by the inner and the outer. And there are the coarsest fibers in the middle of bamboo wall. The fineness of fibers in the outer is comparable to that in the inner.

不同竹子中竹纤维长度和细度随竹竿高度和竹壁厚度的变化分别见表 3-1 和表 3-2。随着竹竿高度的变化，竹竿中部的纤维最长，其次是基部和梢部；竹竿中部的纤维最粗，其次是梢部和基部。随着竹壁厚度的变化，竹壁中部的纤维最长，其次是内部和外部；竹壁中部的纤维最粗，外部和内部纤维粗细相同。

Table 3-1 Variations in length and fineness of bamboo fiber with the height of bamboo stalk
表 3-1 竹纤维长度和细度随竹竿高度的变化

Variety 种类	Length 长度（mm）			Fineness 细度（μm）		
	Root 基部	Middle 中部	Tip 梢部	Root 基部	Middle 中部	Tip 梢部
Phyllostachys edulis 毛竹	1.989	2.055	1.850	15	15	15
Folium bambusae 乌药竹	2.000	2.290	1.530	15.2	13.2	13
Sinocalamus affinis 慈竹	1.940	2.040	1.823	14.8	13.5	14.5

Table 3-2 Changes in length and fineness of bamboo fiber with the thickness of bamboo wall
表 3-2 竹纤维长度和细度随竹壁厚度的变化

Variety 种类	Length 长度（mm）			Fineness 细度（μm）		
	Outer 外部	Middle 中部	Inner 内部	Outer 外部	Middle 中部	Inner 内部
Phyllostachys edulis 毛竹	2.025	2.475	2.250	12.3	14	14.0
Phyllostachys sulphurea 刚竹	1.950	2.325	2.175	15.8	17.2	15.8
Pleioblastus amarus 苦竹	2.100	2.325	2.100	14.0	17.5	14.0

2 Preparation/ 制备

Natural bamboo fiber and bamboo pulp fiber are collectively referred to as bamboo fiber according to the different processing methods. Natural bamboo fiber is a natural cellulose fiber, whereas bamboo pulp fiber is a **regenerated** cellulose fiber, also labelled as "Regenerated bamboo fiber" or "Bamboo viscose fiber".

按照加工方法，竹纤维包括竹原纤维和竹浆纤维。竹原纤维是一种天然纤维素纤维；竹浆纤维是一种再生纤维素纤维，又称再生竹纤维或竹黏胶纤维。

2.1 Preparation of Natural Bamboo Fiber/ 竹原纤维制备

Natural bamboo fiber is obtained directly from

采用机械和物理的方法，从

phyllostachys edulis stalk by mechanical and physical methods. The detailed process is as follows. Depending on the intended use, remove bamboo joints and cut bamboo stalk into chips with certain length, which are suitable to be blended spun with other fibers. **Scour** the bamboo chips in boiling water. Then after scouring process, crush and hammer the chips into filaments. Lignin, hemicellulose, pectin and other impurities in bamboo filaments are removed by **degum**ming. Afterwards, textile fibers are obtained through cleaning, bleaching, oiling, softening, **opening** and **carding**. As shown in Figure 3-3, the preparation process is: Bamboo→Chips→Scouring→Crushing→Degumming→Carding→Textile fibers.

竹竿中直接提取原生纤维而制得竹原纤维。根据用途，去掉竹节后的竹竿剖成一定长度的竹片以满足与其他纤维混纺的需要，将竹片放入沸水中煮练，煮练后取出压碎锤成竹丝，通过脱胶去除竹丝中的木质素、半纤维素和果胶等杂质，然后经清洗、漂白、上油、柔软、开松和梳理即可获得纺织用的竹原纤维。其制作过程（图3-3）为：竹材→制竹片→煮练竹片→压碎分解→脱胶→梳理→纺织用纤维。

Figure 3-3　Preparation process of natural bamboo fiber
图3-3　竹原纤维的制作过程

2.2　Preparation of Bamboo Pulp Fiber/ 竹浆纤维制备

The preparation for bamboo pulp fiber is similar to that for viscose. Any variety of bamboo can be used as raw materials to prepare bamboo pulp fiber by extracting cellulose from bamboo stalk by mechanical and chemical methods and then wet spinning. Through a process of **hydrolysis-alkalization** and multi-stage bleaching, bamboo stalks and leaves are **refine**d into bamboo pulp which could meet the production requirements. Then the bamboo pulp dissolves in sodium hydroxide. After wet spinning and **solidification**, the bamboo pulp is processed into bamboo pulp fiber. The preparation process is: Bamboo pulp → Smashing → Soaking → **Alkaliz**ing → **Sulphona**ting → Pre-dissolving → Dissolving → First **filtration** → Second filtration → **Ripening** → Filtering

竹浆纤维的制作过程基本与黏胶相似。以任意竹材为原料，采用机械和化学方法，从竹竿中提取纤维素再湿法纺丝而制得竹浆纤维。竹竿和竹叶通过水解碱化和多段漂白，精制成可满足生产要求的竹浆粕，经氢氧化钠溶解、湿法纺丝、凝固等工艺即可制得竹浆纤维。其制作过程为：竹浆粕→粉碎→浸渍→碱化→磺化→初溶解→溶解→头道过滤→二道过滤→熟成→纺丝前过滤→纺丝→塑化→水洗→切断→精练→烘干→打包。

before spinning → Spinning → **Plasticiz**ing → Washing → Cutting → Refining → Drying → Packing

Bamboo pulp fiber includes filament and staple fiber (cotton-type or wool-type). Staple fiber can be directly spun into pure yarn, or blended spun with other fibers to achieve blended yarn.

The preparation for natural bamboo fiber is a physical process. It won't cause any pollution to the environment. However, there are certain restrictions in the choice of bamboo as raw materials. The preparation for bamboo pulp fiber is a chemical process that will inevitably pollute the environment. Yet various bamboos are available to be used as raw materials.

In current, most existing bamboo textiles are made up of bamboo pulp fiber blended spun with other fibers, in which the features of bamboo fiber may be more or less damaged. The bamboo fiber in the yarn no longer has the excellent properties of natural bamboo fiber. In many developed countries, this bamboo or bamboo cotton products are not recognized as the bamboo textiles. Bamboo pulp fiber is similar to viscose fiber in property. The wet tenacity of bamboo pulp fiber is low, only half of the dry tenacity. Hence bamboo pulp fiber is not resistant to water **washability** and the size is not stable after washing. It is predicted that natural bamboo textiles will become the mainstream of bamboo textiles in the next period of time.

竹浆纤维包括长丝和棉型、毛型短纤维。短纤维可直接纺成纯纺纱或与其他纤维混纺制成混纺纱。

竹原纤维的制备属物理方法，不造成环境污染，但在竹材原料的选择上有一定的限制。竹浆纤维的制备属化学方法，在制备过程中，不可避免会造成环境污染，但可用各种各样的竹材为原料。

目前，竹纤维纺织品大多是将竹浆纤维与其他纤维混纺制成，该竹制品已经使纤维的特性受到了或多或少的破坏，其纱线中的竹纤维已基本上不再具备竹原纤维的优良特性，许多发达国家已经不再将这种竹或竹棉制品认定为竹纺织品。竹浆纤维性能类似于黏胶纤维，湿强低，约是干强的一半，因此，竹浆纤维不耐水洗，洗后尺寸不稳定。由此可见，竹原纺织品必将成为将来一段时间内的主流竹纺织品。

3 Structure/ 结构

3.1 Molecular Structure/ 分子结构

The chemical composition of bamboo and other **bast** fibers are shown in Table 3–3. Natural bamboo fiber is mainly composed of 45%~55% cellulose, 20%~25% hemicellulose, 20%~30% lignin, 0.5%~1.5% pectin, and

竹原纤维和其他韧皮纤维的化学成分见表3–3。竹原纤维由纤维素、半纤维素、木质素、果胶和其他非纤维素物质

Table 3-3 Chemical compositions of natural bamboo and other bast fibers
表 3-3 竹原和其他韧皮纤维的化学成分

Chemical Compositions 化学成分（%）	Fiber 纤维			
	Natural bamboo 竹原	Hemp 大麻	Flax 亚麻	Jute 黄麻
Cellulose 纤维素	45~55	55~67	70~80	50~60
Hemicellulose 半纤维素	20~25	16~18.5	8~11	12~18
Lignin 木质素	20~30	6.3~9.3	1.5~7	10~15
Pectin substance 果胶物质	0.5~1.5	3.8~6.8	1~4	0.5~1
Water soluble matter 水溶物	7.5~12.5	10~13	1~2	1.5~2.5
Wax 蜡质	—	1~1.2	2~4	0.3~1
Ash 灰分	1.03~1.93	1.6~4.6	0.5~2.5	0.5~1

7.5%~12.5% other noncellulosic substances. The cellulose in natural bamboo fiber is considerably lower than it is in **hemp**, flax, **jute** and other bast fibers. The hemicellulose and lignin in natural bamboo fiber are higher than those in bast fibers, whereas pectin and other components are relatively lower in comparison to those in bast fibers.

Bamboo has been successfully grown in Gansu, Sichuan, Guangdong, Guangxi, Fujian and Hunan Provinces. Diverse varieties and origins lead to different chemical compositions as shown in Table 3-4.

组成，其含量分别为45%~45%、20%~25%、20%~30%、0.5%~1.5% 和 7.5%~12.5%。竹原纤维的纤维素含量明显低于大麻、亚麻和黄麻等韧皮纤维。其半纤维素和木质素含量高于韧皮纤维，果胶等其他成分含量低于韧皮纤维。

在我国，竹子的主要产地为甘肃、四川、广东、广西、福建和湖南，产地和品种不同，竹纤维的化学成分含量也有所差异，见表3-4。

Table 3–4 Chemical compositions of natural bamboo fibers with different varieties and origins
表 3-4 不同品种和产地的竹原纤维的化学成分

Variety 品种	Origin 产地	Chemical Compositioins 化学成分（%）				
		Cellulose 纤维素（%）	Hemicellulose 半纤维素（%）	Lignin 木质素（%）	Pectin 果胶（%）	Ash 灰分（%）
Phyllostachys edulis 毛竹	Gansu 甘肃	46.50	21.56	23.40	—	1.23
Sinocalamus affinis 慈竹	Sichuan 四川	44.25	25.41	31.28	0.87	1.20
Phyllostachys bissetii 白夹竹	Sichuan 四川	44.35	25.41	31.28	0.87	1.20
Green bamboo 绿竹	Guangdong 广东	49.55	17.45	23.00	—	1.78
Phyllostachys iridescins 丹竹	Guangxi 广西	47.88	18.54	23.55	—	1.93
Phyllostachys edulis 毛竹	Fujian 福建	45.50	21.12	30.67	0.70	1.10
Phyllostachys edulis 毛竹	Hunan 湖南	52.57	23.71	26.62	—	1.03

3.2 Supramolecular Structure/ 超分子结构

The X-ray **diffraction(XRD)** patterns of natural bamboo, **ramie,** flax and cotton fibers are shown in Figure 3-4, which all exhibit a typical XRD pattern of crystal cellulose I. The peaks at $2\theta=14.60°$, $16.25°$, $22.47°$ and $33.85°$ correspond to the diffraction peaks of 101, 10$\bar{1}$, 002 and 040, respectively. The diffraction peaks of natural bamboo and ramie fibers have a relatively high resolution because 002 **crystal** surface shows sharp diffraction peak. The crystallinity of natural bamboo fiber (72%) is close to that of ramie fiber, but higher than that of cotton and flax fibers. The crystallinity of bamboo pulp fiber is similar to that of viscose fiber (about 30%). Natural bamboo fiber has a high degree of **microcrystalline** orientation like ramie and flax fibers, and much higher than cotton fiber.

竹原纤维、苎麻、亚麻及棉纤维的 X 射线衍射曲线如图 3-4 所示。它们都呈现出典型的结晶纤维素 I 的 X 射线衍射曲线，$2\theta=14.60$、16.25、22.47 和 33.85 分别对应于 101、10$\bar{1}$、002 和 040 晶面的衍射峰，竹原纤维和苎麻纤维的衍射峰具有较高的分辨率，002 晶面的衍射峰更尖锐。竹原纤维的结晶度为 72%，接近于苎麻纤维，高于棉和亚麻纤维。竹浆纤维的结晶度与黏胶的基本相同，约为 30%。竹原纤维具有较高的微晶取向度，与苎麻、亚麻纤维差异很小，但远大

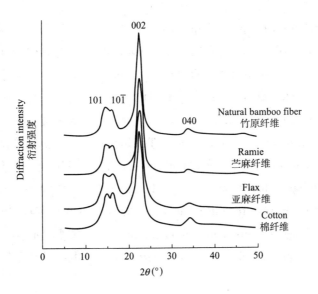

Figure 3-4　X-ray diffraction patterns of fibers
图3-4　纤维的X射线衍射曲线

The microstructure of natural bamboo fiber is shown in Figure 3-5. Natural bamboo fiber has a cross section composed of several concentric layers, namely primary wall (P), secondary wall (S) and lumen. The secondary wall of natural bamboo fiber is much more complicated than that of cotton and ramie fibers. Both cotton and ramie fibers have three layers in the secondary wall. The **microfiber**s

于棉纤维。

竹原纤维的微观结构如图3-5所示。竹原纤维的截面是由许多的同心层组成，可分为初生层（P）、次生层（S）和中腔三个部分。竹原纤维的次生层比棉和苎麻纤维要复杂得多。棉和苎麻纤维的次生层主要分为3层，

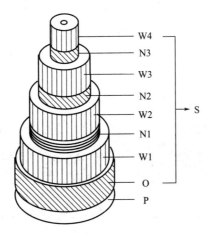

Figure 3-5　Microstructure of natural bamboo fiber
图3-5　竹原纤维的微观结构

P—Primary wall/初生层　O—Outer layer of secondary wall/次生层外层　W1~W4—Wide wall of secondary wall/次生层宽层　N1~N3—Narrow wall of secondary wall/次生层窄层

43

and fiber axis are arranged in spiral shape. The cellulose polymers spiral at 25°~30° in cotton fiber larger than that in ramie fiber (8°~10°). The secondary wall of natural bamboo fiber contains the outer (O) and inner layers. The outer layer is relatively thin, in which the cellulose polymers spiral at 2°~10° in relation to the fiber axis. The inner layer is **alternately** composed of wide (W1~W4) and narrow (N1~N3) layers. The lignin density and the spiral angle in wide layer are lower and smaller than those in narrow layer, especially in the N1 layer. The orientation of microfibril in N1 layer diverges away from the fiber axis, and it gets closer to the **transverse** direction.

微纤与纤维轴均呈螺旋形排列。棉纤维中纤维素大分子倾斜角较大，取向度为25°~30°，而苎麻纤维中，倾斜角小于棉纤维，取向度为8°~10°。竹原纤维次生层包括外层（O）和内层，外层比较薄，呈螺旋形排列，取向度为2°~10°。内层是由宽层（W1~W4）与窄层（N1~N3）交替组合而成。宽层木质素密度较低，倾斜角较小；窄层木质素密度较高，倾斜角较大，特别是N1层，其微纤的取向偏离纤维的轴向而更接近横向。

3.3 Morphological Structure/ 形态结构

Figure 3-6 shows the SEM images of morphology of natural bamboo fiber. As is shown, natural bamboo fiber is slender and long. It is **pointed** at their two ends and **spindle**-shaped. There are innumerous **micro-groove**s on the longitudinal surface. Natural bamboo fiber has transverse segment like flax, but no natural convolution or ribbon shape like cotton fiber. The cross section is irregularly elliptical or kidney-shaped with large lumen. The hollow degree is high. There are many micro-pores with different sizes in cross section and **micro-crack**s around

竹原纤维形态的扫描电子显微镜照片如图3-6所示。竹原纤维细长，两端尖，呈纺锤状，纵向表面有无数微细沟槽。竹原纤维有麻类纤维的横节，但没有棉纤维的天然转曲和带状结构。横截面为不规则的椭圆形、带中腔的腰圆形等，高度中空，横截面上布满了大大小小的孔隙，且边缘有裂纹，与苎麻纤维的横截面

(a) Transverse section of natural bamboo fiber/竹原纤维横向

(b) Longitudinal section of natural bamboo fiber/竹原纤维纵向

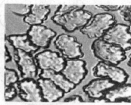
(c) Transverse section of bamboo pulp fiber/竹浆纤维横向

(d) Longitudinal section of bamboo pulp fiber/竹浆纤维纵向

Figure 3-6　SEM images of morphology of bamboo fibers
图3-6　竹纤维形态的扫描电子显微镜照片

the edge, which are similar to those of ramie fiber. Bamboo pulp fiber is similar to viscose fiber in morphology, which has many micro-grooves of different depths along the fiber length, and **serrated** or **petal**-like cross section.

很相似。竹浆纤维形态与黏胶纤维相似，纵向表面有深浅不等的微细沟槽，具有锯齿状或梅花瓣状的横截面。

4 Property/性能

Length and fineness: The length and fineness of bamboo and cotton fibers are shown in Table 3-5. Natural bamboo fiber is longer and coarser than cotton fiber in length and fineness. Bamboo pulp fibers of different lengths and fineness are produced depending on end-use requirements.

长度和细度：竹和棉纤维的长度和细度见表3-5。竹原纤维比棉纤维长、粗，竹浆纤维的长度和细度可根据需要而定。

Table 3-5　Lengths and fineness of bamboo and cotton fibers
表3-5　竹纤维和棉纤维的长度和细度

Fiber 纤维	Natural bamboo fiber 竹原纤维	Bamboo pulp fiber 竹浆纤维	Cotton fiber 棉纤维
Fineness 细度（公支）	1200~2400	1800~6000	4000~8500
Length 长度（mm）	15~100	38~86	23~64

Tensile property: The tensile properties of bamboo pulp and viscose fibers are shown in Table 3-6. The tensile property of bamboo pulp fiber is slightly better than that of viscose fiber. The dry and wet tenacity of bamboo pulp fiber are 2.49 cN/dtex and 1.54 cN/dtex, respectively, which are slightly higher than that of viscose fiber (2.26 cN/dtex when dry and 1.46 cN/dtex when wet). The elongation ratio at breaking for bamboo pulp fiber is 17.5% when dry and 7.0% when wet. Bamboo pulp fiber has a slightly higher initial modulus than viscose fiber. Bamboo pulp fiber therefore has better **processability** and **wearability** than viscose fiber. Both fibers have lower wet tenacity, thus they rank low in resistance to be water laundered.

强伸性：竹浆和黏胶纤维的强伸性见表3-6，竹浆纤维的强伸性略优于黏胶纤维。竹浆和黏胶纤维的干、湿强度分别为 2.49 cN/dtex、1.54 cN/dtex 和 2.26 cN/dtex、1.46 cN/dtex，竹浆纤维的干、湿强度略高于黏胶纤维。竹浆纤维的干、湿断裂伸长率分别为 17.5% 和 7.0%。竹浆纤维具有略高于黏胶纤维的初始模量。这些表明竹浆纤维的可加工性和服用性好于黏胶纤维。竹浆和黏胶纤维的湿强都低于干强，因此，不耐水洗。

Table 3-6 Tensile properties of bamboo pulp and viscose fibers
表 3-6 竹浆和黏胶纤维的强伸性

Fiber 纤维		Strength 强力（cN）	Breaking elongation 断裂伸长（%）	Linear density 线密度（dtex）	Tenacity 强度（cN/dtex）	Initial modulus 初始模量（cN/dtex）	Specific rupture work 断裂比功（J/cm³）
Bamboo pulp fiber 竹浆纤维	Dry 干	3.34	17.50	1.33	2.49	52.35	0.26
	Wet 湿	2.04	7.00	1.31	1.54	34.60	0.06
Viscose fiber 黏胶纤维	Dry 干	3.80	17.90	1.66	2.26	51.50	0.24
	Wet 湿	2.40	7.70	1.63	1.46	31.40	0.06

Moisture absorption and air permeability: As a cellulose fiber with lots of hydrophilic groups, bamboo fiber has good moisture absorption. Bamboo fiber is **hygroscopic** and its moisture regain ranks the top of most fibers. In the standard state the moisture regain of natural bamboo and bamboo pulp fibers are 7.03% and 12.11%, respectively. The crystallinity in natural bamboo fiber is considerably larger than that in bamboo pulp fiber. Bamboo pulp fiber therefore is better than natural bamboo fiber in terms of moisture absorption. The moisture absorption curve of bamboo pulp fiber at 95% relative humidity and 20 ℃ is shown in Figure 3-7. As is shown, the moisture regain of bamboo pulp fiber can reach as high as 45%. And the

吸湿透气性：竹纤维属于纤维素纤维，具有较多的亲水性基团，吸湿性好。竹纤维是高吸湿纤维，其回潮率居大多数纤维之首。在标准状态下，竹原纤维和竹浆纤维的回潮率分别为7.03%和12.11%。竹原纤维的结晶度远大于竹浆纤维，因此，竹浆纤维的吸湿性好于竹原纤维。竹浆纤维在温度20 ℃和相对湿度95%时的吸湿曲线如图3-7所示。竹浆纤维在温度为20 ℃、相对湿度为95%的回潮率高达

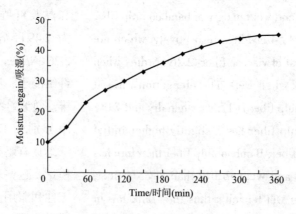

Figure 3-7 Moisture absorption curve of bamboo pulp fiber
图3-7 竹浆纤维的吸湿曲线

moisture absorption rate is particularly fast. It takes only 6 hours for the moisture regain to increase from 8.75% to 45%.

Natural bamboo fiber is noted for its air permeability, which is in the first place among all the fibers. This is partially due to the innumerous micro-grooves along the fiber surface. It is also due to lots of micro-pores with different sizes arranged **uniformly** in the highly hollow cross section. All these contribute to the instant **absorbency** and quick evaporation of water. The moisture absorption and air permeability of natural bamboo fabric are 3.5 times those of cotton fabric.

Dyeability: The chemical composition of bamboo pulp fiber is similar to that of cotton and viscose fibers, therefore the **dyestuff** for cotton and viscose fibers can also be used for bamboo pulp fiber. Bamboo pulp fiber has better dyeability than viscose fiber. Several reasons contribute to its dyeability. Low crystallinity and large amorphous area allow the dyestuff to enter into the fiber interior. Besides, high moisture absorption and high wet **swell**ing ratio increase the **diffusion** coefficient in fiber and the dyeing ratio.

Natural antibacteria: Bamboo fiber exhibits many desirable properties such as antibiosis, bacteriostasis, anti-mite, **anti-insect** and **anti-odor** as a result of natural antibacterial substance "bamboo **quinone**". The bamboo quinone keeps active after the fiber is constantly washed and exposed to sunlight, and causes no **allergy** to human skin. Table 3-7 shows the **steriliz**ing ratios of cellulose fibers. It is shown that bamboo, **linen** and ramie fibers all have a comparatively strong antibacterial effect, much better than cotton fiber. The sterilizing effect of bamboo fiber is shown in Figure 3-8. It could be seen that 48% and 75% harmful bacteria are killed in 1 hour and 24 hours, respectively.

Deodorization and adsorption: Natural bamboo

45%，且吸湿速率特别快，回潮率从8.75%增加到45%仅需6 h。

竹原纤维纵向表面的无数微细沟槽以及高度中空的横截面上均布满了大大小小的孔隙，可以在瞬间吸收并蒸发水分，因此，竹原纤维的透气性居于纤维之首。竹原织物的吸湿透气性是棉织物的3.5倍。

染色性：竹浆纤维的化学组成与棉、黏胶纤维相同，用于棉、黏胶纤维的染料同样能用于竹浆纤维染色。竹浆纤维染色性好于黏胶纤维。这是由于竹浆纤维结晶度较低，无定形区大，有利于染料进入纤维内部；另外，竹浆纤维吸湿性强，在湿态下膨胀率高，可提高染料在纤维内的扩散系数和上染率。

天然抗菌性：由于含有天然抗菌物质"竹醌"，竹纤维具有天然抗菌、抑菌、防螨、防虫和防异味等优异功能，经反复洗涤和暴晒后，"竹醌"仍能保持原有的活性，而且不会对人体皮肤造成任何过敏性反应。纤维素纤维的杀菌率见表3-7，竹、亚麻、苎麻纤维均具有较强的抗菌作用，比棉纤维好得多。竹纤维的杀菌效果如图3-8所示。1 h和24 h后，48%和75%有害细菌被杀死。

除臭和吸附性：竹原纤维具

Table 3-7 Sterilizing ratios of cellulose fibers
表 3-7 纤维素纤维的杀菌率

Strain 菌种	Sterilizing ratio 杀菌率（%）				
	Natural bamboo fiber 竹原纤维	Ramie fiber 苎麻纤维	Flax fiber 亚麻纤维	Cotton fiber 棉纤维	Bamboo pulp fiber 竹浆纤维
Staphylococcus aureus 金黄色葡萄球菌	99.0	98.7	93.9	—	94.9
Bacillus subtilis 枯草芽孢杆菌	99.7	98.3	99.8		53.8
Candida albicans 白色念珠菌	94.1	99.8	99.6	40.1	85.1
Aspergillus niger 黑曲霉菌	83.0	51.2	—	—	—

(a) Harmful bacterium/
有害菌

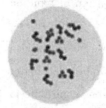

(b) Kill 48% harmful
bacterium after 1 hour/
1 h后杀灭48%有害菌

(c) Kill 75% harmful
bacterium after 24 hours/
24 h后杀灭75%有害菌

Figure 3-8 Sterilizing effect of bamboo fiber
图3-8 竹纤维的杀菌效果

fiber has fairly good deodorization and adsorption. The presence of **sodium chlorophyll** and the special superfine microporous structure enable the fiber to evaporate the absorbed harmful substances when exposure to the sunlight, which automatically leads to the functional regeneration of deodorization and adsorption of the fiber. Natural bamboo fabric forcefully **adsorb**s harmful substances such as **formaldehyde**, **benzene**, **toluene**, **ammonia**. The deodorization ratio of natural bamboo fabric to ammonia and acid reach about 70%~72% and 93%~95%, respectively, which are exceptionally higher than those of cotton fabric.

Ultraviolet resistance: Natural bamboo fiber is born with excellent ultraviolet resistance owing to the existence of sodium chlorophyll, a safe and excellent UV absorber.

有良好的除臭和吸附性，由于含有叶绿素铜钠以及特殊的超细微孔结构，该结构能在阳光下蒸发吸收的有害物质，从而实现除臭和吸附功能的自动再生。竹原纤维织物能强烈吸附甲醛、苯、甲苯和氨等有害物质。竹原纤维织物对氨气和酸的除臭率分别为70%~72%和93%~95%，除臭效果比棉织物要好得多。

防紫外性：叶绿素铜钠是安全、优良的紫外线吸收剂，因而竹原纤维具有良好的、与生俱来

The ultraviolet transmittance for cotton fiber is about 2500/10000, which approximately equal to 417 times that for natural bamboo fiber (6/1000). The ultraviolet resistance ratio of natural bamboo fabric is up to 98%. In addition, the **anion** produced by natural bamboo fiber could also block the ultraviolet effectively from human body. Therefore, natural bamboo fabric is a natural shield for human body and causes no **irritation** to skin.

Softness and comfort: Through a process of **degreas**ing, **deacidification** and **deproteinization**, bamboo fabric feels soft and comfortable, and will not harden after successive use.

Biodegradablity: Bamboo fiber is susceptible to **decompose** into **carbon dioxide** and water by the action of **microorganism** and sunshine when buried in soil or activated **sludge**.

In addition, bamboo fabric is **lustrous** and exhibits excellent behaviors such as **drapability**, **stiffness** and resilience.

的防紫外性。竹原纤维的紫外线透过率为6/10000，棉纤维的紫外线透过率为2500/10000，是竹原纤维的417倍。竹原纤维织物的防紫外线率高达98%。另外，竹原纤维能产生负离子，可以有效阻挡紫外线对人体的辐射，是人体的天然屏障，不会对皮肤产生任何刺激。

柔软舒适性：进行脱脂、脱酸和脱蛋白处理后，竹织物手感柔软舒适，连续使用后不会发硬。

可生物降解性：埋入土壤或活性污泥中，在微生物和阳光的作用下，竹纤维可分解成对环境无任何污染的二氧化碳和水。

此外，竹织物色泽亮丽，具有良好的悬垂性、硬挺性和回弹性。

5 Application/ 应用

Both natural bamboo fiber and bamboo pulp fiber have possibly met the spinning requirements considering their fineness, length, tenacity and other physical and mechanical properties. A variety of pure and blended bamboo fiber products have been introduced to the mass market so far, such as apparel, home textiles, **sanitary** and medical materials, etc.

Apparel and home textiles: Bamboo fiber is processed into fabrics such as towel, bath towel, underwear, T shirt, sportswear, summer wear, summer fabric due to its excellent moisture absorption, air permeability, soft and comfortable **feel**. It is also made into socks owning to its

无论是竹原纤维还是竹浆纤维，其细度、长度和强度等物理机械性能均能满足纺纱的要求。目前，市场已有各种纯纺和混纺竹纤维产品，用于服装、家纺和卫生医疗领域。

服装家纺面料：由于竹纤维吸湿透气性强，手感柔软、舒适，适合做毛巾、浴巾、内衣、T恤、运动服和夏季服装及面料。由于其除臭性，适合做袜

deodorization. In addition, it is suitable to be applied in swimming suit due to its anti-ultraviolet.

Sanitary and medical materials: Bamboo fiber has natural **sterilization** and bacteriostasis. It therefore has been extensively used as sanitary materials such as pads, facial masks, food packages, etc, and medical materials such as masks, bandages, surgical clothes, nurse wears, etc.

卫生医疗材料：由于竹纤维具有天然的抗菌和抑菌性，已广泛用作卫生巾、面膜和食品包装等卫生材料，口罩、绷带、手术服和护士服等医疗材料。

6 Problem/ 问题

There are two difficulties in preparing natural bamboo fiber so far. One is that the single fiber is too short to be spun. The other is that the lignin content in natural bamboo fiber is too high to be removed. The conventional process of chemical degumming is long and complicated. Lots of energies are supposed to be consumed and the equipments are severely corroded during the process, leading to the serious environment pollution. And the quality of the output is proved to be unstable. The biological degumming method is not easily implemented either because of the tight structure and high density of bamboo combined with large amount of air in the fiber cell. It's hard for the impregnation solution to **penetrate** into natural bomboo fiber, leading to the extension of the degumming time. Besides, too many antibacterial substances in bamboo complicate the choice of bacteria. As to the bamboo pulp fiber, the prolonged process not only deteriorates the environment, but also the properties of raw bamboo could easily be destroyed during the process. Further research and exploration need to be conducted on the process of bamboo fiber.

In addition, bamboo fiber tends to break off in weaving process due to the strong moisture absorption, high wet elongation ratio and large plastic deformation. Other properties like handle, drapability and water shrinkage need to be improved as well.

目前，竹原纤维的制备主要有两个难点：一是单纤维太短，无法纺纱；二是竹原纤维中的木质素含量很高，难以去除。常规的化学脱胶方法工艺流程长，工艺复杂，需消耗大量的能量，且设备腐蚀严重，环境污染极为严重，加工出来的纤维品质不稳定。而生物脱胶法实施难度大，竹材自身结构紧密，密度很大，而且纤维细胞中又有大量空气存在，浸渍液很难浸透，脱胶时间延长。且竹子本身具有多种抑菌物质，菌种的选择较困难。竹浆纤维生产工艺过程过长，对环境污染严重，竹材原料特性在加工过程中容易被破坏。因此，竹纤维的加工工艺有待进一步研究和探索。

另外，由于竹纤维吸湿强、湿伸长以及塑性变形大的特点，在织造过程中易断。竹织物的手感、悬垂性及缩水性等有待改善。

Exercises/ 练习

1. Name classification and definition of bamboo fibers.
2. Describe the properties and applications of bamboo fibers.
3. Compare the properties of bamboo fabrics with the properties of cotton fabrics.
4. Translate the following Chinese into English.

（1）竹纤维就是从自然生长的竹子中提取出的一种纤维素纤维，是继棉、麻、毛、丝之后的第五大天然纤维。

（2）竹纤维是一种天然环保型绿色纤维。被称为"会呼吸"的纤维和"纤维皇后"。

（3）竹纤维具有良好的透气性、瞬间吸水性、耐磨性和染色性等特性，同时，又具有天然抗菌、抑菌、除螨、防臭和抗紫外线功能。

（4）竹纤维的吸湿透气能力是棉的3.5倍，是由于高度中空的横截面布满了大大小小椭圆形的孔隙，可以在瞬间吸收并蒸发大量的水分。

（5）竹纤维的抗紫外线能力是棉的417倍。

References/ 参考文献

[1] 李宁, 白洋. 竹纤维性质及其应用 [J]. 纺织科技进展, 2007（3）: 15-17.

[2] 郭豫吉, 邓首哲. 竹纤维性能及发展展望 [J]. 中国麻叶科学, 2008, 30（6）: 321-325.

[3] 程隆棣, 徐小丽, 劳继红. 竹纤维的结构形态及性能分析 [J]. 纺织导报, 2003（5）: 101-103.

[4] 周衡书, 钟文燕. 竹纤维的开发与应用 [J]. 纺织科学研究, 2003（4）: 30-26.

[5] 何建新, 章伟, 王善元. 竹纤维的结构分析 [J]. 纺织学报, 2008, 2（29）: 20-24.

[6] 隋淑英, 李汝勤. 竹纤维的结构与性能研究 [J]. 纺织学报, 2003, 24（6）: 535-536.

[7] 曹泰钧, 刘刚毅. 竹纤维纺织品开发及前景 [J]. 湖南文理学院学报: 自然科学版, 2005, 17（1）: 57-59.

[8] 范杰. 竹纤一种新型的纺织纤维材料 [J]. 天津纺织科技, 2005（2）: 35-37.

[9] 黄大兵. 竹纤维在纺织品领域的应用 [J]. 中国纤检, 2011（9）: 78-79.

Part Two

Protein Fiber/ 蛋白质纤维

Chapter Four   Milk Protein Fiber/ 牛奶蛋白纤维

1   Introduction/ 前言

Milk protein fiber is a chemical fiber containing animal protein. According to its manufacturing method or chemical component, milk protein fiber is divided into pure milk fiber and composite milk fiber. Pure milk protein fiber is a regenerated fiber obtained by extracting protein from milk and then wet spinning. Composite milk fiber is a **semi-synthetic** fiber which is obtained by **copolymerizing** the protein extracted from milk with polymer and then **wet spinning**. At present, milk protein fiber available is mainly semi-synthetic fiber containing milk **casein** and **polyacrylonitrile** or **polyvinyl alcohol**. It is a new animal-protein fiber also called milk silk or milk fiber, which is different from natural fiber, regenerated fiber and synthetic fiber. It exhibits both the advantageous properties of natural fiber and synthetic fiber, which not only satisfies people's pursuit of wearing comfort and **aesthetic**, but also is in line with the trend toward health and **fitness**.

Milk fiber was first made in 1935 by SNIA, an Italian company, and Coutaulds Corporation from England, experiments have been carried on in an effort to produce a more delicate and luxurious silk-like fiber. However, only the 2.9% protein in milk was utilized as raw material. The manufactured fiber not only cost high, but also had remarkably lower tenacity. There was a big gap between

牛奶蛋白纤维是一种含动物蛋白质的化学纤维，根据生产方式或化学成分，牛奶蛋白纤维分为纯牛奶纤维和复合牛奶纤维。纯牛奶纤维是一种从牛奶中提取蛋白质，再经湿法纺丝而得到的再生纤维。复合牛奶纤维是一种从牛奶中提取的蛋白质与高聚物共聚，再经湿法纺丝而得到的半合成纤维。目前，牛奶蛋白纤维主要是含乳酪蛋白和聚丙烯腈或聚乙烯醇的半合成纤维。它是一种有别于天然纤维、再生纤维和合成纤维的新型动物蛋白纤维，又称牛奶丝、牛奶纤维。它集天然纤维和合成纤维的优点于一身，满足了人们对穿着舒适性、美观性的追求，符合保健、健康的潮流。

为了人工制造出更精美、更高档的仿真丝纤维，1935年，意大利的SNIA公司和英国的Coutaulds公司仅利用牛乳中2.9%的蛋白质作为制取牛奶纤维的原料，不仅成本高，而且制造出的纤维强度太低，性能与天

milk protein fiber and silk in terms of properties, and there was little use value. In 1956, with the development of synthetic fiber, a new type of milk silk with silk-like structure was invented by **graft copolymerization** of milk protein solution and polyacrylonitrile in Japan TOYOBO Corporation. The fiber was first put into industrial production and marketed under the **trademark** Chino in 1969, targeting for **surgical suture**.

Research on milk fiber was not conducted until 1960s in China. Due to the complicated technological process, limited technical condition and short milk supply, milk can not be selected as the only raw material for milk fiber. Milk fiber was successfully produced by Zhengjia Milk Filament Technology Co., Ltd in Shanghai during 1990s. In 1995, "Chemical Properties and Identification Methods of Milk Fiber" was formulated by Three Gun Group in Shanghai and the company became the first national testing organization capable of identifying milk fiber in China. In July, 2002, Shanghai Entry-Exit Inspection and Quarantine Bureau upgraded this method to the industry standard SHCIQH 0003—2001 "qualitative test method for milk fiber textile", which became the only qualitative test method for milk fiber by domestic testing institutions. Later, this industry standard was upgraded to the national standard.

然真丝差距很大，没有使用价值。随着合成纤维的发展，1956年日本东洋纺公司用牛乳蛋白溶液与聚丙烯接枝共聚开发了类似于真丝结构的新型牛奶长丝，1969年首次实现工业化生产，商品名为Chino，产品定位是手术缝合线。

20世纪60年代，我国才开始研究牛奶纤维。由于制造纤维的工艺流程复杂，技术条件有限，牛乳供应紧缺，不能只采用牛乳作为牛奶纤维的原材料。90年代，上海正家牛奶丝科技有限公司成功研制出牛奶纤维。1995年，上海三枪集团制定了《牛奶纤维化学性能和鉴别方法》，成为我国最早有能力鉴别牛奶纤维的国家检测机构。2002年7月，上海市出入境检验检疫局将该方法升级为行业标准：SHCIQH 0003—2001《牛奶纤维纺织品定性检验方法》，该标准成为国内检测机构对牛奶纤维唯一定性检验方法。后来，该行业标准升级为国家标准。

2　Preparation/ 制备

2.1　Principle/ 原理

The protein in milk is capable to form fiber because it satisfies the primary conditions of being fiber-forming polymer.

(1) **Macromolecules** in milk are **linear.** There are two

牛奶中的蛋白质之所以能制成纤维，是因为它具备成纤高聚物的基本条件。

（1）大分子是线型的。有两

kinds of protein macromolecular structures. One is linear chain described as **fibrin** (fibrous protein) and the other is **globular** chain described as **globulin**. The fibrin can be found in milk and made into fiber. The globulin is found in **soybean** and cannot be made into fiber.

(2) Macromolecules have favorable **flexibility** and **intermolecular** forces. The presence of innumerable **peptide linkage**s in protein accounts for the excellent flexibility of the macromolecules. Consequently **hydrogen bond**s can be formed among adjacent peptide linkages, resulting in high intermolecular forces. Apart from that, other **polar group**s such as —SH, —PH$_2$, also make a contribution to high intermolecular forces.

(3) The protein polymer in milk has good spinnability. Protein is soluble in water to form **colloidal solution**. After spinning the macromolecules approach to each other to form hydrogen bonds with the removal of the water. The **polypeptide** chains are arranged in parallel or even twisted together to be **solidified** into water insoluble **strands**. The tenacity of the strands can be up to 2.5 cN/dex, which satisfies the basic requirements of textile fiber.

种蛋白质大分子结构，一种是线型链，称作纤维蛋白；另一种是球状链，称作球蛋白。牛奶中的蛋白质是线状的，可以成纤；而大豆中的蛋白质是球状的，则不能成纤。

（2）大分子具有一定的柔性和分子间作用力。蛋白质分子中含有无数个肽键，使得大分子具有很好的柔性，因此，相邻肽键间形成氢键，从而使其具有较大的分子间作用力。此外，其他极性基团如—SH、—PH$_2$等对分子间作用力也有辅助作用。

（3）具有较好的可纺性。蛋白质与水形成胶体溶液，经纺丝后，随着水分的去除，大分子相互靠拢，分子间形成氢键，多肽链平行排列，甚至扭在一起，固化为不溶于水的丝条。丝条的强度可达到 2.5 cN/dex 以上，能满足纺织纤维的基本要求。

2.2 Preparation Technology/ 制备工艺

2.2.1 Pure Milk Fiber/ 纯牛奶纤维

Water content represents more than 85% in milk. The initial step in the production of pure milk fiber is to remove excess water to **concentrate** the water less than 60%, and then the fat in milk is removed as well. Finally, the protein molecules and small molecular substances in the solution are **isolate**d by **dialysis** or **salting out** to obtain **purifi**ed protein. Pure milk fiber production process is: evaporating→degreasing→alkalizing→isolating→kneading→filtrating→deaerating→spinning→**stretch**ing→drying→setting

牛奶中水分占85%以上，制备纯牛奶纤维，首先去除水分，浓缩到含水60%以下，然后去除脂肪。最后，通过透析法或盐析法分离溶液中蛋白质分子和小分子物质，从而纯化蛋白质。纯牛奶纤维生产工艺流程为：蒸发→脱脂→碱化→分离→揉合→过滤→脱泡→纺丝→拉伸→

grading → packing.

Decompressing evaporation: Water content in milk should be concentrated to less than 60% under 70 ℃ and 80 kPa, because protein is susceptible to lose its activity above 70 ℃.

Degreasing: Milk is suggested to be physically degreased before alkalization to reduce the burden of alkalization, alkali consumption and the production cost. A large proportion of fat could be removed by using a **centrifuge** in respect that fat has lower density than other components in milk. The speed of centrifuge is required to be higher than 1000 r/min.

Alkalization: Alkali chemical degreasing method is taken for further fat removal. After **centrifugal** degreasing, NaOH solution is added into the **emulsion** to break down the fat completely. The optimum mole ratio of alkali and water is proved to be 1∶20. If alkali is insufficient, the fat won't be thoroughly broken down. Conversely, the excessive alkali may cause protein to break down as well.

Isolating: The protein is obtained by **filter**ing the alkalized solution through **semipermeable membrane**. It also could be separated out by adding **inorganic salt**s like **magnesium sulfate (MgSO$_4$), sodium chloride (NaCl)**, etc into the solution. A chemical reaction known as salting out takes place to **precipitate** the protein out.

Kneading: The protein dissolves in a mixture of **deionized** water and protein adhesive to form protein solution. It is heated to 60 ℃ in a kneading machine and sufficiently mixed to obtain spinning solution.

Filtration and deaeration: Filtration and deaeration remove impurities and bubbles in spinning solution, respectively. The **pilling and broken ends** are subsequently reduced, which improves the spinnability.

干燥→定型→分级→包装。

减压蒸发：在温度低于70 ℃、压力低于80 kPa条件下，将牛奶浓缩到含水60%以下，因为蛋白质在70 ℃以上会失去活性。

脱脂：为减少碱化负担，降低碱耗量，节约生产成本，在碱化前，先进行物理脱脂。脂肪的密度小于牛奶中其他成分的密度，利用离心机将牛奶中大部分脂肪去除。离心机转速要求高于1000 r/min。

碱化：为进一步去除脂肪，利用碱进行化学脱脂。在离心脱脂后的乳液中加入氢氧化钠溶液，使脂肪完全分解。碱液中碱和水的最佳摩尔比为1∶20。碱量太少，脂肪不能完全分解；量太多，多余的碱会使蛋白质分解。

分离：采用半透膜过滤碱化后的溶液得到蛋白质，或者在溶液中加入硫酸镁（MgSO$_4$）和氯化钠（NaCl）等无机盐析出蛋白质。发生了一种叫作盐析的化学反应，使蛋白质沉淀出来。

揉合：将蛋白质溶解于去离子水中，并加入蛋白质黏合剂形成蛋白溶液，在揉合机中，加热至60 ℃，经充分混匀而得到纺丝液。

过滤和脱泡：分别去除纺丝液中的杂质和气泡，减少纺丝中的毛丝、断头，提高可纺性。

Spinning: Dry spinning method is generally adopted. The technological process is:

$$\text{Spinning solution} \xrightarrow[\text{Preheating}]{60\ ℃} \text{Spinning can} \longrightarrow$$

$$\text{Spinneret} \xrightarrow[\text{Air solidifying}]{55\ ℃} \text{As spun fiber}$$

$$\xrightarrow[\text{Drying}]{100\ ℃} \text{Winding}$$

The filtered and deaerated spinning solution is heated to 60℃ and placed into spinning can. The solution is then forced through spinnerets (**orifice** diameter of 0.15 mm) by the use of **gear metering pump** and pulled in fine streams. These fine streams are initially solidified into **as-spun fiber**s in hot air at 55 ℃, which pass through the drying area at 100 ℃. The water content decreases to less than 20%. They are **wind**ed on the **cone**.

Stretching: The **nascent fiber** should be thermally stretched 15~20 times to improve the breaking tenacity and other physical and mechanical properties of the fiber. After stretching, macromolecules are **align**ed along the fiber axis and the intermolecular force is enhanced, which helps to improve the properties of the fiber.

Drying and setting: The fiber is dried and **heat-set** at 90 ℃ under vacuum to reduce the moisture content by less than 10%. After heat treatment, the protein is converted into **denatured protein**, thus the fiber is **permanent**ly non-melting and insoluble, which is resistant to hot water and softening, and becomes a valuable textile fiber.

2.2.2 Composite Milk Fiber/ 复合牛奶纤维

Composite milk fiber has the similar spinning technique to pure milk fiber. There are three ways to prepare the spinning solution: blending, **crosslinking** and graft copolymerization. The production process of composite

纺丝：采用干法纺丝方法，其工艺流程为：

$$\text{纺丝液} \xrightarrow[\text{预热}]{60\ ℃} \text{纺丝罐} \longrightarrow$$

$$\text{喷丝头} \xrightarrow[\text{空气固化}]{55\ ℃} \text{初生纤维}$$

$$\xrightarrow[\text{烘干}]{100\ ℃} \text{卷绕}$$

将过滤和脱泡后的纺丝液加热至60℃，置于纺丝罐中，经齿轮计量泵输入纺丝机的喷丝头（喷丝孔径为0.15 mm）。从喷丝孔挤出形成细流后，进入温度约为55℃的热空气中初步固化形成初生纤维，再经过100℃的烘干区，使水分含量降低至20%以下，并卷绕在筒子上。

牵伸：为提高纤维的断裂强度及其他物理机械性能，成形后的初生纤维还需要进行15~20倍的热拉伸。拉伸后，大分子沿纤维轴取向排列，分子间作用力增强，纤维性能得到改善。

干燥和定型：纤维在90℃真空下进行干燥和定型，使纤维含水率低于10%。经热处理后，蛋白质转化为变性蛋白质，成为永久的不熔不溶性固化纤维，耐热水，耐软化，成为有实用价值的纺织纤维。

复合牛奶纤维和纯牛奶纤维的纺丝方法相同。复合牛奶纤维的纺丝液制备有三种方法，即共混法、交联法、接枝共聚法。复

milk fiber is shown in Figure 4-1.

合牛奶纤维生产工艺流程如图 4-1 所示。

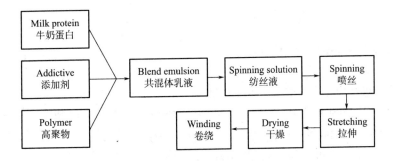

Figure 4-1 Production process of composite milk fiber
图4-1 复合牛奶纤维生产流程

Blending: The milk protein is blended with polyacrylonitrile to obtain the spinning solution. The preparation is simple and no chemical reaction ever occurs. However, the milk protein is poorly **dispersed** in the form of cylindrical **aggregation** with the diameter of 30~50 nm and the length of 100 nm. The **uneven** dispersion affects the quality of milk fiber. The excellent properties of milk fiber will be lost.

Crosslinking: Milk protein and polyacrylonitrile occur to crosslinking reaction under the action of crosslinking agent to obtain the spinning solution. Milk protein is **evenly** dispersed in the form of nanoparticles less than 20 nm in diameter.

Graft copolymerization: Graft copolymerization takes place under the action of **catalyst** to obtain the spinning solution when milk protein is blended with polyacrylonitrile. Milk protein is uniformly scattered in the form of molecules in polyacrylonitrile. The obtained fiber enjoys the best quality, but the preparation of spinning solution is rather complicated.

共混法即牛奶蛋白质和聚丙烯腈共混而制得纺丝液。制备方法简单，没有任何化学反应，但牛奶蛋白质分散较差，主要以直径为 30~50 nm，长度为 100 nm 的圆柱状凝聚体分散，分散的不均匀性，影响牛奶纤维质量，失去牛奶纤维的优良性能。

交联法即牛奶蛋白质和聚丙烯腈在交联剂作用下发生高聚物交联反应而制得纺丝液。牛奶蛋白质分散均匀，分散颗粒直径小于 20 nm。

接枝共聚法即牛奶蛋白质和聚丙烯腈混和，在催化剂作用下发生高聚物接枝共聚反应而制得纺丝液。牛奶蛋白质以分子状均匀地分散在聚丙烯腈中，此法得到的纤维质量最好，但纺丝原液的制备工艺比较复杂。

3 Structure/ 结构

3.1 Chemical Composition/ 化学组成

Composite milk fiber is chemically composed of milk protein and polyacrylonitrile. They constitute the **crystalline** region (70%) and amorphous region (30%), respectively, which is close to those of silk. The crystalline and amorphous sections occupy 80% and 20% in silk, respectively. Casein is the polymer protein composed of diverse **amino acid**s linked by **peptide bond**s. The composition of amino acids is complex and **dispersed**. Most amino acids have relatively larger **side group**s. Therefore, the polymers are not prone to **crystallize**. The crystallinity and crystalline **grain size** in composite milk fiber and other fibers are calculated by the peak-splitting method. The results are shown in Table 4-1. Composite milk fiber has close crystallinity and crystalline grain size with polyacrylonitrile fiber, but crystallinity is lower in composite milk fiber than that in silk.

复合牛奶纤维的化学成分为牛奶蛋白和聚丙烯腈，分别构成结晶区和无定形区，各占70%和30%。这与真丝比较接近，真丝中结晶区和无定形区各占80%和20%。酪素是多种氨基酸以肽键相连的高分子蛋白质，其氨基酸组成复杂、分散，大部分具有较大的侧基，因此不易结晶。复合牛奶纤维和其他纤维的结晶度和晶粒尺寸用分峰法计算后结果见表4-1，其结晶度和晶粒尺寸与聚丙烯腈纤维非常接近，结晶度低于蚕丝纤维。

Table 4-1 Crystal structure parameters of composite milk fiber and other fibers
表 4-1 复合牛奶纤维和其他纤维结晶结构参数

Fiber 纤维	Crystallinity 结晶度（%）	Crystal size L_{100} 晶粒尺寸（nm）
Composite milk fiber 复合牛奶纤维	70	4.54
Polyacrylonitrile fiber 聚丙烯腈纤维	73	4.52
Silk 蚕丝纤维	80	—

3.2 Morphological Structure/ 形态结构

Figure 4-2 shows the SEM images of the cross section and surface morphology of milk fiber. The cross section

牛奶纤维的横截面及表面形态的扫描电子显微镜照片如图

with micro-pores is flat, **dumbbell** or kidney-shaped. It is **profiled fiber**. There are irregular micro-grooves and island-shaped **concave and convex** on the longitudinal surface, which result from the fiber surface **dehydration** during the spinning process and fast orientation in the fiber. Besides, milk fiber has a **curly**, milky white or **yellowish** appearance.

4-2所示。横截面呈扁平状，呈哑铃形或腰圆形，布有细小的微孔，属于异形纤维。纵向表面有不规则的沟槽和海岛状的凹凸，这是由于纺丝过程中纤维的表面脱水、纤维快速取向形成的。另外，牛奶纤维具有一定的卷曲，外观呈乳白色或微黄色。

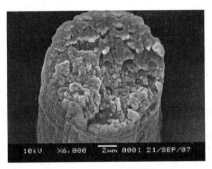
(a) Cross section/横截面

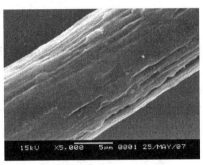
(b) surfacemorpholoyy/表面形态

Figure 4-2 SEM images of milk fiber
图4-2 牛奶纤维的扫描电子显微镜照片

4 Property/ 性能

4.1 Tensile Property/ 强伸性

Table 4-2 presents the breaking tenacity of milk fiber and silk. The wet tenacity of milk fiber is lower than dry tenacity but the difference in dry and wet tenacity is smaller than that of silk, which indicates that the moisture content of milk fiber has some influence on its tensile property, but it is not significant. Hence **humidification** is generally applied to remove static electricity during spinning.

牛奶纤维和蚕丝的断裂强度见表4-2。牛奶纤维的湿态强度低于干态强度，但干湿态强度的差异小于蚕丝，这说明牛奶纤维的含水率对其拉伸性能有一定的影响，但不显著，因此，在纺纱过程中，可以采用加湿的方法去除静电。

Table 4-2 Breaking tenacity of milk fiber and silk
表 4-2 牛奶纤维和蚕丝的断裂强度

Fiber 纤维	Tenacity 强度（cN/dtex）	
	Dry 干	Wet 湿
Milk fiber 牛奶纤维	2.8~5.33	2.4~3.8
Silk 蚕丝	3.8~5.7	1.9~2.8

4.2 Moisture Absorption and Conductivity/ 吸湿导湿性

Milk fiber has a moisture regain of 5%~8%, lower than silk (8%~9%) despite that milk fiber has larger amorphous region than silk. This is because the moisture absorption is determined by fiber microstructure and hydrophilic groups. It is difficult for water molecules to penetrate into the interior of fiber crystalline region. But the crystalline surface plays an important role in moisture absorption. The crystalline part of silk is a polypeptide chain molecule with high hydrophilicity, which easily attracts a large number of water molecules.

Milk protein molecules, which contain natural protein **moisturiz**ing factor and a large number of hydrophilic groups, distribute on the fiber surface. Sweat generated by the wearer could be instantly absorbed and expelled from the micro-grooves to the air, resulting in dry skin all the time. Milk fiber therefore has excellent moisture absorption and permeability.

虽然牛奶纤维的无定形区大于蚕丝，但牛奶纤维的回潮率为5%~8%，低于蚕丝的回潮率8%~9%。这是因为纤维吸湿性是由纤维的微观结构和亲水基团共同决定的。水分子很难渗透到纤维结晶部分的内部，但结晶表面对吸湿起很大作用，蚕丝结晶部分是亲水性较大的肽链分子，易吸水。

牛奶蛋白质分子含有天然蛋白保湿因子和大量亲水基团，分布在纤维的表面，可迅速吸收人体汗液，通过沟槽快速导入空气中散发，使肌肤始终保持干爽状态。因此，牛奶纤维具有良好吸湿透湿性能。

4.3 Friction Property/ 摩擦性

As an important surface property, friction property influences fiber cohesion, yarn formation, abrasion, deformation of the fiber, and the product handle as well. Its friction property is due to its characteristic surface

纤维的摩擦性能是一项重要的表面性能，影响纤维的抱合、成纱、磨损和变形以及成品的手感等。牛奶纤维的摩擦性与其特

morphological structure. The non-circular cross section and micro-grooves along the longitudinal direction play a decisive role in it. Owing to its high **friction coefficient**, as well as to the excellent cohesion among fibers, **crimp**ing process is not necessary for milk fiber when it is made by mechanical method. And the spun yarns have relatively high tenacity, which is conducive to spinning.

有的表面形态结构有关，非圆形的横截面和纵向表面的沟槽对纤维的摩擦性能起决定性作用。牛奶纤维的摩擦系数较高，纤维间的抱合力好，因此，在采用机械方法加工时，可以不进行卷曲加工，而且纤维成纱后，其强度较高，有利于纺纱加工。

4.4 Electrical Property/ 电学性

Milk fiber has higher **mass specific resistance** than cotton fiber but lower than silk and polyacrylonitrile fiber. Oil and antistatic agents are highly recommended to be added during the spinning process in case of the serious static discharge (Table 4-3).

牛奶纤维的质量比电阻高于棉纤维，但低于蚕丝和聚丙烯腈纤维，因此，在纺纱加工过程中静电现象仍比较严重，须加适当的油剂和抗静电剂（表 4-3）。

Table 4-3 Mass specific resistance of milk and other fibers
表 4-3 牛奶纤维和其他纤维的质量比电阻

Fiber 纤维	Milk 牛奶	Cotton 棉	Silk 蚕丝	**Acrylic fiber** 腈纶
Mass specific resistance 质量比电阻（$\Omega \cdot g/cm^2$）	1.8×10^9	$10^6 \sim 10^7$	$10^9 \sim 10^{10}$	$10^{13} \sim 10^{14}$

4.5 Dyeability/ 染色性

Milk fiber has excellent dyeability. It is considered to have high dye-up ratio by **cationic**, acid and **reactive dye**s. The dyed products are bright and lustrous. They are resistant to degradation by sun and sweat with the color fastness up to Level 4. The contradiction of bright dyeing color and color fastness is successfully solved in milk fiber compared with silk product.

牛奶纤维具有较好的染色性。可以用阳离子、酸性和活性染料染色，上染率高，颜色鲜艳而有光泽，日晒和汗渍色牢度高达 4 级以上，与真丝产品相比，牛奶纤维产品成功解决了染色鲜艳与色牢度的矛盾。

4.6 Comfort/ 舒适性

Milk fiber has equivalent moisture absorption with cotton fiber, and far better moisture conductivity and air

牛奶纤维的吸湿性与棉纤维相当，而其导湿透气性远优于棉

permeability than cotton fiber. The crimp recovery ratio of milk fiber is close to that of cashmere and wool. Thus milk fiber is fine, soft and **fluffy**, and has cashmere-like feel. Milk fiber exhibits warm in winter and cool in summer behavior due to its **three-dimensional** and micro-porous structure in the cross section and micro-grooves in the longitudinal direction. Light-weight fabric made of milk fiber quickly absorbs the sweat from skin, which fast spreads to the air to keep the skin cool in summer. Micro-pores in thick fabric absorb the heat emitted by the body, which effectively obstructs air circulation and prevents cold air from entering in winter. It is light and has warmth retention. Therefore, milk fiber has better comfort.

4.7 Aesthetics/ 外观

Milk fabric has silk-like luster and excellent drapability. The level of **anti-fuzzing** and **anti-pilling** is up to 3–4.

4.8 Light Resistance/ 耐光性

After continuous exposure to ultraviolet, the tenacity of milk fiber decreases little, which indicates that it has good light resistance.

4.9 Healthcare Function/ 保健功能

Milk fiber has good skin affinity due to a variety of essential amino acids. Many hydrophilic groups, such as —NH$_2$, —OH and —COOH in protein macromolecules, also have excellent moisturizing effect on skin. In addition, natural bacteriostasis and antibacterial ratio of milk fiber reach up to 99% and 80%, respectively. Hence, milk fiber has a good and lasting **healthcare** function.

好持久的保健功能。

5 Fiber Identification/ 纤维鉴别

According to FZ/Y 01057.4—99, milk fiber is identified by alkali dissolving method. Sample is immersed in 2.5% NaOH solution, which is heated for 30 min at constant temperature of 100 ℃. If fiber gradually swells and forms **gel**, and the color becomes dark red from milky white and then fades away into **buff** (pale yellow), it is the milk fiber.

根据 FZ/Y 01057.4—99，采用碱溶解法进行牛奶纤维的鉴别。将试样浸泡在 2.5% 氢氧化钠溶液中，在 100 ℃ 恒温条件下，加热 30 min，纤维逐渐溶胀形成凝胶，颜色从奶白色先变成深红色，然后褪成淡黄色，即为牛奶纤维。

6 Problem/ 问题

Although milk fiber has many excellent properties due to its unique structures, there are still some problems that need to be further solved.

牛奶纤维因其独特结构而有许多优良的特性，但它还存在一些问题，有待进一步完善。

6.1 Poor Heat Resistance/ 耐热性差

Milk fiber **yellow**s above 120 ℃ and browns above 150℃ when placed in dry and hot condition. If placed in **moist** thermal condition for over 1 hour, it slightly turns yellow below 100 ℃. It is suggested that the laundering temperature should not exceed 30 ℃ and ironing temperature should be within 120 ℃. It is best to iron it at the low temperature (80~120 ℃).

牛奶纤维在干热状态下，120 ℃ 以上泛黄，150 ℃ 以上变褐色；在湿热状态下，100 ℃ 以下保持 1h 以上轻微泛黄。因此，洗涤温度不要超过 30 ℃，熨烫温度不要超过 120 ℃，最好使用"低温"（80~120 ℃）熨烫。

6.2 Low Chemical Stability/ 化学稳定性低

Milk fiber has relatively low alkali resistance like other protein fibers. **Chloride** bleaching agent should be avoided.

牛奶纤维的耐碱性与其他蛋白质纤维类似，具有较低的耐碱性。不能用氯化物漂白剂漂白。

6.3　Color/ 颜色

Pure white milk fiber products are not available in respect that the light yellow pigment is not removable.

因为牛奶纤维的淡黄色无法去除，所以不能得到纯白色的牛奶纤维产品。

6.4　High Price/ 价格高

Milk fiber and its products are pricy because the manufacturing technique of milk fiber is not mature yet and it is impossible to realize **large-scale** and **batch production** resulting in the very low output.

目前，牛奶纤维的生产工艺还不是很成熟，无法实现大规模、批量化生产，因此产量很低，所以牛奶纤维及其产品的价格相对较高。

7　Application/ 应用

So far, the primary applications of milk fiber are in the following areas.

目前，牛奶纤维主要应用在以下几个方面。

7.1　Yarns/ 纱线类

Milk fiber mainly includes milk filaments of 83.33 dtex, 111 dtex and 166.65 dtex, and milk protein staple fibers of 1.67 dtex × 38 mm, which could be purely spun or blended spun with other fibers such as cashmere, wool, silk, Tencel, cotton, modal, etc.

主要有 83.33 dtex、111 dtex 和 166.65 dtex 牛奶长丝，1.67 dtex × 38 mm 牛奶蛋白短纤维，可制成牛奶纯纺纱线以及与羊绒、羊毛、蚕丝、天丝、棉和莫代尔等其他纤维混纺制成混纺纱线。

7.2　Woven and Knitted Fabrics/ 机织和针织面料

So far there are mainly four series of milk fabrics in market, Pw, Lm, Sm and Am. Pw and Lm are knitted plain and **ribbed fabric**s made of pure milk yarns, which are suitable for casual and household apparel, such as T-shirts and **underwear** for men and women, etc. Sm is woven

主要有 Pw、Lm、Sm 和 Am 四大系列牛奶纤维面料。Pw 和 Lm 是以纯纺牛奶纱制成的针织平纹和罗纹面料，适合制作男女 T 恤、内衣等休闲家居服装。Sm fabric made of milk/silk blended yarns. The fabric exhibits the characteristics of milk fiber, which is light, smooth and drapable, and the behavior of silk, which has flexibility and **toughness**, and bright and **gorgeous** style. It is suitable for Tang suit, **cheongsam**, evening dress and other high-grade **fashion**. Am is **elastic** fabric made of milk/**polyurethane** blended yarn. It has many favorable features such as softness and resilience, which is suitable for knitted **blazer**, rhythm **workout clothes** and body-building underwear.

是以牛奶/蚕丝混纺纱制成的机织面料,集牛奶纤维和真丝的优点于一身,既有牛奶纤维轻盈、爽滑、悬垂性好的特性,又具有真丝柔中带韧、光洁艳丽的风格,适宜制作唐装、旗袍、晚礼服等高级服装。Am是以牛奶/氨纶混纺纱制成的弹力面料,具有柔软、弹性适度的优点,适合制作针织运动上衣、韵律健身服和美体内衣。

7.3 Home Textiles/ 家纺类

Home textiles, made of blended yarn of milk fiber and other fibers, have fine-grained, **breathable** and smooth **texture**, **luxurious** luster and bright color. The quilt filled with milk fiber is fluffy and soft. It has good **heat preservation**, **elasticity**, sleep-promoting function, anti-mite and antibacteria, which is especially suitable for **allergic** people.

以牛奶纤维与其他纤维混纺纱制得的家纺面料,质地细密轻盈,透气爽滑,光泽优雅华贵,色彩艳丽。以牛奶纤维为填充物制成的牛奶被松软,保温性能良好且富有弹性,具有促进睡眠、防螨抗菌功能,特别适用于过敏体质的人群。

7.4 Nonwoven Fabrics/ 非织造布

Nonwovens made of milk fiber are extensively applied in medical and health fields. The products contain **bandages**, **gauzes**, **pads**, baby and adult **diapers**, and external application materials that are fluffy, soft, comfortable, antibacterial and **deodorant**. It also has regulatory effect on skin and no **tickle** to skin.

牛奶纤维制成的非织造布产品广泛应用于医疗和卫生领域,主要有绷带、纱布、妇女卫生巾、婴儿和成人尿不湿、外用敷贴材料等产品。产品蓬松、柔软舒适、抑菌防臭,对皮肤有调理作用,无刺痒感。

Exercises/ 练习

1. How to classify milk protein fiber?
2. Why protein in milk can be directly spun into fiber?
3. Describe the milk protein fiber preparation process.
4. How to identify the milk protein fiber?

References/ 参考文献

[1] 蔡忠波, 朱军军, 刘优娜, 等. 牛奶蛋白纤维的发展与展望 [J]. 中国纤检, 2015, 4: 82-84.

[2] 冯建永. 牛奶蛋白纤维的结构、性能及应用 [J]. 化纤与纺织技术, 2008, 12: 27-29.

[3] 李克兢, 何建新, 崔世忠. 牛奶蛋白纤维的结构与性能 [J], 纺织学报, 2006, 27 (8): 57-60.

[4] 王自强, 成玲. 牛奶蛋白纤维的生产及开发应用 [J]. 纺织科技进展, 2008, (1): 45-47.

[5] 杨乐芳. 牛奶蛋白纤维的生态功能性评价及产品开发 [J]. 上海纺织科技, 2007, 35 (2): 46-48.

[6] 郑宇, 程隆棣. 牛奶蛋白纤维的特性、应用和定性检测 [J]. 上海纺织科技, 2006, 34 (6): 56-57.

[7] 张建英, 马晶, 张建波. 牛奶蛋白纤维的物理化学性能 [J]. 印染助剂, 2009, 26 (6): 47-52.

[8] 董勤霞, 潘玉明, 柯华. 牛奶蛋白纤维的性能及其染整加工 [J]. 印染, 2006, 1: 30-33.

Chapter Five   Soybean Protein Fiber/ 大豆蛋白纤维

1 Introduction/ 前言

The main components of soybean include crude fat, crude protein, **polysaccharide**, crude fiber, water and ash. After extracting oil from soybean, the obtained crude fat is soybean oil and the **remains** is **soybean meal**. Soybean protein fiber is a chemical fiber, which is obtained by modifying **globular protein** (**glycinin**) extracted from soybean meal as the raw material, copolymerizing the modified glycinin with polyvinyl alcohol, and then wet spinning.

Soybean fiber is independently developed in China, and it is the first to realize industrial production in the world. It is also the only fiber invention with complete intellectual property right that has been achieved in our country so far. In 1993 Guanqi Li in Huakang Biological Chemical Engineering United Group Ltd began to develop soybean fiber. In 1998 soybean fiber was trial-spun successfully on laboratory equipment. Later on an industrial production line of 1500 tons per year was built. After production **debug**ging for more than 8 months industrial spinning began in March 2000, which rewrote the blank history of Chinese original technology in the field of chemical fiber manufacturing in the world.

大豆的主要成分包括粗脂肪、粗蛋白质、多糖、粗纤维、水和灰分。大豆榨油后，所得的粗脂肪为豆油，剩余的物质为豆粕。大豆蛋白纤维是以豆粕为原料，从中提取球蛋白质，改性后与聚乙烯醇共聚，经湿法纺丝而成的化学纤维。

大豆纤维是由我国自主研发，并在国际上率先实现了工业化生产，也是迄今为止我国获得的唯一完全知识产权的纤维发明。1993年，华康生物化学工程联合集团公司李官奇开始研发大豆纤维，1998年，在实验室的试验设备上试纺成功，然后建成一条年产1500吨的工业化生产线，经8个多月的生产调试，在2000年3月开始进行工业化纺丝，改写了在世界化学纤维制造领域中中国原创技术空白的历史。

2 Preparation/ 制备

Soybean protein molecule contains very few **nonpolar** amino acids with flexible molecular chains such as **imidic acid**, **lactamic acid**, **cysteine** and **cystine** and many polar amino acids with larger side groups. It is easy to form globular molecular chain instead of linear molecular chain. Soybean protein therefore is a **spherical** protein, which does not meet the conditions of fiber polymer. It can not directly be spun into fiber.

The production process of soybean fiber is shown in Figure 5-1. The soybean meal is soaked in water and separated, and the soybean globulin is extracted. Chemical reagents are applied to change the **spatial** structure into

大豆蛋白质分子中柔性链非极性氨基酸特别是乙氨酸、丙氨酸、半胱氨酸和胱氨酸等含量极少，具有较大侧基的极性氨基酸含量较高，大豆蛋白质分子很难形成线型分子链，而呈球状分子链，因此，大豆蛋白质为球状蛋白质，不满足成纤高聚物的条件，不能直接纺丝而成纤维。

大豆纤维生产工艺流程如图 5-1 所示。将豆粕浸水和分离，提取出大豆球蛋白质，利用化学试剂将其空间结构改变成直链结

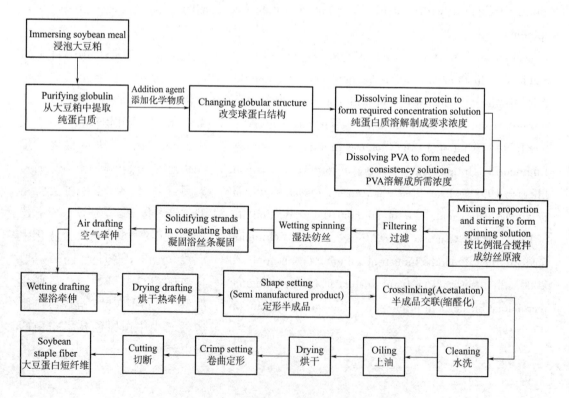

Figure 5-1 Production process of soybean fiber
图5-1 大豆纤维的生产工艺流程

straight chain structure, and the soybean linear protein is then formed. Soybean linear protein and polyvinyl alcohol (PVA) dissolve to form 15%~20% solution, respectively. Protein solution and PVA solution are mixed in proportion and **stir**red to form spinning solution for wet spinning.

After filtering and deaeration, the spinning solution is extruded through a spinneret to **coagulation** bath to form nascent fiber, which is then subjected to hot air, wet and dry **draft**ings to form soybean filament. Soybean staple fiber is produced by **acetalization**, water washing, oiling, drying, **curl**ing and crimp-setting and cutting.

构，从而形成大豆线型蛋白质。分别溶解大豆线型蛋白质和聚乙烯醇形成15%~20%的溶液，把蛋白质溶液和聚乙烯醇溶液按配比混合并搅拌均匀制成纺丝液，进行湿法纺丝制备大豆纤维。

纺丝液过滤、脱泡后，从喷丝孔挤出进入凝固浴形成初生纤维，经热空气牵伸、湿牵伸和干牵伸三步牵伸后，形成大豆长丝，再经缩醛化、水洗、上油、烘干、卷曲定型和切断，形成大豆短纤维。

3 Structure/ 结构

3.1 Chemical Structure/ 化学结构

Soybean fiber is composed of 25%~45% protein and 75%~55% polyvinyl alcohol (PVA). PVA contains a large number of **hydroxyl** groups. Protein constitutes hydroxyl, amino, **carboxyl** and **sulfur**-containing groups, etc. These polar groups form hydrogen bonds, **salt bond**s and **disulfide bond**s among macromolecules. In spinning, it is possible to produce chemical bonds with higher energy, such as **amide** bonds and **ester** bonds. In **acetalation**, the presence of formaldehyde, which is involved in the chemical reaction, is responsible for the intermolecular **crosslinkage** among macromolecules. Therefore, the aggregation structure of soybean fiber is the linear macromolecule **network structure**.

大豆纤维由25%~45%的蛋白质和75%~55%的聚乙烯醇组成。聚乙烯醇含有大量羟基，蛋白质含有较多的羟基、氨基、羧基和含硫基等，这些极性基团在大分子间形成氢键、盐式键和双硫键。在纺丝中，可能产生能量较高的化学键，如酰胺键和酯键等。在缩醛化中，甲醛参与了化学反应，大分子间形成了分子间交联。因此，大豆纤维的聚集态结构是直线型大分子网状结构。

3.2 Morphological Structure/ 形态结构

SEM images of cross section and surface morphology of soybean fiber are shown in Figure 5-2. The cross section of soybean fiber is kidney-shaped or dumbbell-shaped. There are some micro-grooves on the unsmooth longitudinal surface. Besides, soybean fiber is generally crimped, but the **crimpness** is not as obvious as that of fine wool. Soybean fiber generally is pale yellow in appearance.

大豆纤维的横截面及表面形态的扫描电子显微镜照片如图5-2所示。截面呈腰圆形或哑铃形，不光滑的纵向表面有沟槽。此外，大豆纤维具有一定的卷曲，但卷曲度不如细羊毛明显，其外观一般呈淡黄色。

(a) Cross section/横截面

(b) Surface morphology/表面形态

Figure 5-2　SEM images of soybean fiber
图5-2　大豆纤维的扫描电子显微镜照片

4 Property/ 性能

4.1 Mechanical Property/ 机械性能

The mechanical properties of soybean and several other fibers are shown in Table 5-1. In comparison, soybean fiber is considerably higher in tenacity. It has remarkably higher dry tenacity than other four fibers. The wet tenacity for soybean fiber is moderately lower than that for cotton fiber but higher than that for other three fibers, which shows that soybean fiber exhibits the similar behavior to viscose fiber in that the tenacity decreases dramatically after moisture

大豆与其他几种纤维的机械性能见表5-1。大豆纤维的强度较高，干态强度比其他四种纤维高得多，湿态强度略低于棉，但高于另外三种纤维，说明大豆纤维吸湿后，与黏胶纤维类似，强度下降明显。因此，在纺纱过程中应适当地控制其含湿量，以保

Table 5-1 Mechanical properties of soybean and other fibers
表 5-1 大豆纤维和其他纤维的机械性能

Property 性能		Soybean fiber 大豆	Cotton 棉	Wool 毛	Silk 丝	Viscose 黏胶纤维
Tenacity 强度（cN/dtex）	Dry 干	4.2~5.4	3.0~4.9	1.0~1.7	3.4~4	1.7~2.3
	Wet 湿	3.9~4.3	3.3~6.4	0.76~1.63	2.1~2.8	0.8~1.2
Elongation ratio 伸长率（%）	Dry 干	18~21	37	25~35	15~24	10~24
	Wet 湿	21	—	25~50	27~33	24~35
Initial modulus 初始模量（cN/dtex）		20~85	68~93	12~25	80~100	65~85
Loop strength 钩接强力（cN）		75~85	70	80	60~80	30~65
Knot strength 打结强力（cN）		85	90~100	85	80~85	45~60
Friction coefficient 摩擦系数	Static 静态	0.235	0.22	0.20~0.25	0.52	0.43
	Dynamic 动态	0.287	0.29~0.57	0.38~0.49	0.26	0.19

absorption. The moisture content therefore is suggested to be properly controlled to ensure the smooth spinning process.

Soybean fiber has higher breaking elongation ratio when wet than dry, which is closely related to supramolecular structures (crystallization, orientation, intermolecular force, etc.) in the fiber. It is advised to distinguish the difference in dry and wet properties in the process design.

Soybean fiber has relatively high loop strength and low knot strength. It is important to determine the **hook strength** and knot strength of single fiber for understanding knitting **coil** fastness and fiber fastness at the knot point.

Initial modulus reflects the ability of resisting deformation when a fiber is subjected to low tensile strength. The initial modulus of soybean fiber has a larger range,

证纺纱过程的顺利进行。

纤维的断裂伸长率与其超分子结构（结晶度、取向度、分子作用力等）密切相关，大豆纤维的湿态断裂伸长率大于干态断裂伸长率，因此，工艺设计时应注意干、湿态性能的差异。

大豆纤维的钩接强力高，打结强力相对偏低。测定单纤钩结强力和打结强力对了解针织线圈牢度和纤维打结处牢度有重要意义。

初始模量反映纤维受较小拉伸力时抵抗变形的能力。大豆纤维的初始模量值变化范围较大，

indicating high strength **unevenness**, which may bring unfavorable effect on spinning and weaving.

Fiber friction property not only directly affects the spinning process, but also affects the quality of yarns and fabrics. The friction coefficient of soybean fiber is lower than that of other fibers and the difference between the dynamic and static friction coefficient is not obvious, which results in the poor cohesive force among fibers. The spun yarns are loose and are prone to break. A certain amount of oil is recommended to be added in spinning process. Soybean fiber is suitable for the products with soft **touch** due to its small friction coefficient.

说明其强力不匀率高，会给纺纱和织造带来不利影响。

纤维的摩擦性不仅直接影响纺纱过程的顺利进行，而且还关系到纱线和织物的质量。大豆纤维的摩擦系数相对其他纤维偏低，且动、静摩擦系数差值较小，使纤维间抱合力差，纺出的纱线松散、易断，纺纱过程中应加入一定量的油剂。由于其摩擦系数小，大豆纤维适合加工手感柔软的产品。

4.2　Crimp Property/ 卷曲性

The crimp property of soybean and other fibers is presented in Table 5-2. The smooth fiber surface results in poor cohesion among fibers, which may bring difficulties to textile processing. Crimp-setting is proved to improve spinnability and elasticity of fiber, and **bulkiness** of fiber assembly, which endows the fabrics with many outstanding performances such as softness, **plumpness**, **crease resistance** and warmth retention. Besides, crimp-setting also improves the luster of fiber and fabric to some extent.

大豆和其他纤维的卷曲性见表 5-2。纤维表面光滑，纤维间抱合力差，给纺织加工带来困难，加上一定的卷曲后，可以提高纤维的可纺性和弹性及纤维集合体的膨松性，使织物柔软丰满，具有良好的抗皱性和保暖性。同时，卷曲对改善纤维和织物的光泽也有一定作用。

Table 5-2　Crimp property of soybean and other fibers
表 5-2　大豆和其他纤维的卷曲性

Fiber 纤维		Soybean 大豆	Cotton 棉	Wool 毛	Silk 丝	Viscose 黏胶纤维
Crimp number 卷曲数（个/cm）		5.2	—	6~9	—	4.8~5.6
Crimp elasticity 卷曲弹性	Crimp ratio 卷曲率（%）	1.65	—	—	—	—
	Residual crimp ratio 残余卷曲率（%）	0.88	—	—	—	—
	Elastic recovery ratio 弹性回复率（%）	72	74	99	54~55	99

The number of fiber crimp is determined by friction force and cohesion among fibers. Too many crimps cause excessive cohesion among fibers, which results in the generation of electrostatic phenomena and damage to fibers. Fewer crimps may lead to poor fiber cohesion, which affects textile processing and decreases yarn quality. The friction coefficient of soybean fiber is smaller and the fiber cohesion is lower. Soybean fiber therefore is crimp-set by mechanical method. The crimp of soybean fibers is obviously reduced after being subjected to force, which indicates that the cohesion among fibers will decrease during spinning process and affect the yarn strength. The fiber elastic recovery ratio of crimp is related to its stiffness. Soybean fiber has low resilience ratio due to its small initial modulus.

纤维间的摩擦力和抱合力决定的纤维的卷曲数，卷曲数过多会使纤维间抱合力过大，产生静电现象和纤维损伤；卷曲数少，纤维间抱合力低，影响纺织加工和成纱质量。大豆纤维的摩擦系数较小，纤维间抱合力较小，因此，用机械方法加上一定的卷曲。大豆纤维受力后卷曲明显降低，说明纺纱过程中纤维间的抱合力会减小，影响成纱的强力。纤维的卷曲弹性回复率与它的刚度有关，大豆纤维的初始模量较小，因而弹性回复率也低。

4.3 Antistatic Property/ 抗静电性

The mass specific resistance of soybean and other fibers is summarized in Table 5-3. As is shown, the mass specific resistance of soybean fiber is very similar to that of silk fiber. It is not easy to generate static electricity in processing, which shows soybean fiber therefore has good antistatic property that is beneficial to processing and wearing.

大豆纤维与其他几种纤维的质量比电阻见表5-3。大豆纤维的质量比电阻接近于蚕丝，在加工中不易产生静电。这表明该纤维的抗静电性比较好，对加工和服用有利。

Table 5-3 Mass specific resistance of soybean and other fibers
表 5-3 大豆和其他纤维的质量比电阻

Fiber 纤维	Soybean 大豆	Cotton 棉	Wool 毛	Silk 丝	Viscose 黏胶纤维
Mass specific resistance 质量比电阻（$\Omega \cdot g/cm^2$）	2×10^9	$10^6 \sim 10^7$	$10^8 \sim 10^9$	$10^9 \sim 10^{10}$	10^7

4.4 Optical Property/ 光学性能

The optical and other physical properties of the fiber are **anisotropic** due to the difference in fiber macromolecule orientation along the fiber axis. The **birefractive index** of soybean and other fibers is shown in Table 5-4. As is

纤维大分子沿纤维轴向的取向差异，使纤维的光学性质和其他物理性质呈现各向异性。大豆纤维和其他纤维的光学双折射率

shown, soybean fiber has lower **birefringence** than other fibers due to its low orientation degree. Apart from that, its high moisture regain also affects the refractive index.

见表5-4。大豆纤维的双折射率低于其他纤维，因为大豆纤维的取向度低，此外，纤维的回潮率较高，也影响了光的折射率。

Table 5-4 Birefractive index of soybean and other fibers
表5-4 大豆和其他纤维的双折射率

Fiber 纤维	Soybean 大豆	Cotton 棉	Wool 毛	Silk 丝	Viscose 黏胶纤维
Birefractive index 双折射率	0.002	0.046	0.010	0.053	0.020

4.5 Heat Resistance/ 耐热性

Soybean fiber is noted for poor heat resistance. It shrinks dramatically over 100 ℃, and its handle is **rough** and **stiff** after cooling. The thermal shrinkage ratio of soybean fiber in dry hot air or boiling water is higher than that of wool, which indicates the poor wet thermal stability. The breaking tenacity and elongation ratio of soybean fiber at 150 ℃ for 30 min are 68% and 92.5% of the initial values, respectively. The processing temperature therefore should be below 110 ℃. Heat setting or dyeing at high temperature is not recommended.

大豆纤维以耐热性较差著称，超过100℃收缩明显，冷却后手感粗糙硬挺。大豆纤维在干热空气或沸水中的热收缩率均高于羊毛，说明其湿热尺寸稳定性差。大豆纤维在150℃、30 min下的断裂强度和断裂伸长率分别为初始值的68%和92.5%，因此，加工温度应低于110℃，不适宜高温热定型和高温染色。

4.6 Appearance and Handle/ 外观与手感

Soybean fabric has an elegant and soft luster, soft and smooth feel, light-weight and good drapability, which are similar to silk/cashmere blended fabric.

大豆织物的外观与手感类似于真丝与羊绒混纺织物，光泽优雅柔和，手感柔软滑爽，质地轻薄，具有良好的悬垂性。

4.7 Moisture Absorption and Fast Drying/ 吸湿快干性

It is noted for moisture absorption partly because the fiber has a dumbbell-shaped or kidney-shaped cross section, micro-grooves along the fiber length, a large

大豆纤维吸湿性优良，一定程度上是因为大豆纤维截面呈腰圆形或哑铃形，纵向有沟槽，含

number of hydrophilic groups such as hydroxyl, amino, carboxyl, etc, large amorphous region and more micro-cracks and micro-holes in fiber. Large specific surface area is conducive to moisture conductivity. Soybean fabric exhibits good moisture absorption and fast drying.

有大量的羟基、氨基、羧基等亲水基团，非结晶区较大，纤维内部有较多的缝隙和孔洞；比表面积大，有利于导湿，因此，大豆纤维织物具有良好的吸湿快干性能。

4.8 Dyeability/ 染色性

Soybean fiber ranks high in acid resistance and can be dyed with acid dyes, reactive dyes and **direct dye**s. The soybean fabric dyed with reactive dyes is bright and lustrous, and is better than silk fabric in color fastness.

大豆纤维耐酸，可用酸性、活性和直接染料进行染色。活性染料染色的大豆纤维织物鲜艳而有光泽，染色牢度优于真丝织物。

4.9 Antibacterial Activity/ 抗菌性

The antibacterial effect is considered to be significantly lasting when the Chinese herbal medicine with **antiseptic** and **anti-inflammatory** effect is added in the spinning process. It reacts with protein side chain to form chemical bond. Soybean fiber has bacteriostasis on escherichia coli, staphylococcus aureus and candida albicans.

在纺丝过程中，加入具有杀菌消炎作用的中草药与蛋白质侧链以化学键相结合，抗菌效果显著持久。大豆纤维对大肠杆菌、金黄色葡萄球菌、白色念珠菌均有抑菌作用。

4.10 Healthcare Function/ 保健功能

Soybean fiber has good healthcare function. The fiber is composed of amino acids which are necessary for human body, and the contained amino group **therein** is **compatible** with human skin. The formaldehyde content in soybean fiber is far below the domestic and foreign standards on formaldehyde content in textiles. Soybean fabric therefore does not cause any harm to human body.

大豆纤维含有人体所必需的氨基酸，含有的氨基与人体皮肤相容，因此，具有良好的保健作用。大豆纤维中的甲醛含量大大低于国内外相关法规对纺织品甲醛含量的限定标准，因此，大豆纤维织物不会对人体造成伤害。

4.11 Comfort/ 舒适性

Soybean fiber has a relatively good heat-moisture

大豆纤维具有良好的热湿舒

comfort. The higher thermal resistance provides better warmth retention for soybean fiber than it is for cotton and Tencel. Besides, soybean fiber is noted for good moisture absorption because it contains a large number of amino, carboxyl and other hydrophilic groups. The moisture regain of soybean fiber is close to that of cotton fiber. Moisture conductivity and air permeability of soybean fiber are superior to those of cotton fiber due to the micro-grooves on soybean fiber surface.

适性。大豆纤维的热阻较大，保暖性优于棉和天丝。此外，大豆纤维含有大量的氨基、羧基等亲水基团，使其具有良好的吸湿性，回潮率与棉纤维接近。大豆纤维表面的沟槽，使其导湿透气性远好于棉。

4.12 Chemical Resistance/ 耐化学性

Soybean fiber contains amino and carboxyl groups, which can absorb both acid and alkali. In strong acid, the strength damage of soybean fiber are 5.5% and 19.2%, respectively, after it is treated for 60 min at pH=1.7 and at pH=11. It can be seen that the acid resistance of soybean fiber is better than alkali resistance. Soybean fiber has good **oxidation-reduction** resistance. The **reductant sodium thiosulfate** ($Na_2S_2O_4$) and the oxidant **sodium hypochlorite** (NaOCl) and **hydrogen peroxide** (H_2O_2) are recommended as bleaching agents to bleach soybean fiber.

大豆纤维中含有氨基（—NH_2）和羧基（—COOH），既可吸收酸也可吸收碱。在强酸中，pH=1.7，处理 60 min 后强力损伤为 5.5%；pH=11 处理 60min 后强力损伤为 19.2%，可见大豆纤维的耐酸性比耐碱性好。大豆纤维有较好的耐氧化还原性，可选择还原剂 $Na_2S_2O_4$ 及氧化剂 NaOCl 和 H_2O_2 作为漂白剂，对大豆纤维进行漂白。

5　Application/ 应用

Soybean fiber exhibits many advantages of both natural and chemical fibers, such as cashmere-like soft feel, wool-like warmth retention, cotton-like moisture absorption and air permeability, silk-like luster, chemical fiber-like moisture conductivity and fast dry, etc. It can be blended spun with various fibers including cotton, wool, silk, viscose, polyester, and polyamide and so on. It is an ideal material for high-grade textile fabric.

大豆纤维兼具天然纤维和化学纤维的诸多优点，如羊绒般的柔软手感、羊毛般的保暖性、棉般的吸湿透气性、真丝般的光泽和化学纤维般的导湿快干性等，可与棉、羊毛、蚕丝、黏胶纤维、涤纶和锦纶等多种纤维混纺，是高档纺织面料的理想材料。

Knitted underwear and sleepcoat: The underwear made of fine soybean fiber is soft, smooth and comfortable. The outer layer of soybean fiber is mainly protein, which has health care effect on human skin. In spinning, Chinese herbal medicine is grafted on protein macromolecule so that the fiber has significant bactericidal and anti-inflammatory effects and keeps good moisture absorption and air permeability. Soybean fiber therefore has a great market potential in the underwear field.

Shirt fabric: Soybean woven fabric has the style of flax/silk blended product in luster but stiffer handle than silk. It is an ideal material for making high-grade shirt due to its good drapability and better **wrinkle resistance** than those of silk.

Blended fabric: Soybean fiber is usually blended spun with silk, wool, cashmere and other fibers. When soybean fiber is blended spun with silk, the wet and cold feeling of silk fabric generated by wet sweat clinging to body can be avoided due to the excellent moisture conductivity of soybean fiber. When it is blended spun with wool, the obtained products achieve excellent performance like wool fabric and the cost is significantly reduced. When it is blended spun with cationic polyester fiber, the comfort of polyester product is proved to be improved. With the continuous development of soybean fiber product in the future, the many advantages, such as light weight, softness, smoothness, high tenacity, moisture absorption, moisture conductivity and air permeability, can give more unique styles to blended products.

针织内衣和睡衣：大豆纤维较细，制作的内衣柔软、光滑和舒适。大豆纤维外层主要为蛋白质，对人体皮肤具有保健作用。在纺丝时，在蛋白质大分子上接枝中草药成分，从而使该纤维具有显著的杀菌消炎的功效，并保持良好的吸湿透气性，因此，大豆纤维在内衣领域内大有市场潜力。

衬衫面料：大豆机织物在光泽上具有麻绢混纺产品的风格，手感比绢硬挺，悬垂性好，抗皱性优于真丝，是制作高档衬衫的理想材料。

混纺面料：大豆纤维可以与蚕丝、羊毛、山羊绒等纤维混纺。与蚕丝混纺时，利用大豆纤维优良的导湿性，可以避免真丝织物由于汗湿而紧贴在身上所产生的湿冷感；与羊毛混纺制得织物可获得羊毛织物的优良性能，还可以显著降低生产成本；与阳离子涤纶混纺，可改善涤纶产品的舒适性能。将来随着大豆纤维产品的不断开发，利用其质轻、柔软、光滑、高强、吸湿、导湿和透气等优点，可以赋予混纺产品更多独特的风格。

Exercises/ 练习

1. What's the definition of soybean protein fiber?
2. Why is the soybean protein molecule globular?

3. What excellent properties do the fabrics made of soybean protein fibers have?

4. Compare soybean protein fiber and milk protein fiber.

5. Translate Chinese into English

（1）大豆蛋白纤维有羊绒般的柔软手感，蚕丝般的柔和光泽，棉的保暖性和良好的亲肤性。

（2）大豆蛋白纤维有明显的抗菌功能，被誉为"新世纪的健康舒适纤维"。

（3）大豆蛋白质纤维制成的单丝，细度细、强伸度高、耐酸耐碱、吸湿性好。

（4）用精梳纱织成的织物，表面纹路清晰，是高档的衬衣面料。

（5）大豆蛋白纤维与各种天然纤维混纺可大量应用在家用纺织品面料领域。

References/ 参考文献

[1] WANG L M, SHEN Y, DING Y, et al. Dyeing performance of soybean fiber treated with low temperature plasma [J]. Journal of Donghua University : English Edition, 2006, 23（4）: 120-122.

[2] 吴佩云. 大豆蛋白纤维和牛奶蛋白纤维的鉴别方法 [J]. 上海纺织科技, 2009, 37（12）: 49-52.

[3] 崔红. 大豆蛋白纤维性能及其产品开发 [D]. 天津: 天津工业大学, 2003.

[4] 宋心远. 大豆纤维的结构性能与染整加工研究 [D]. 上海: 东华大学, 2006.

[5] CHOI J H, KANG M J, YOONB C. Dyeing properties of soya fibre with reactive and acid dyes [J]. Coloration Technology, 2005, 121: 81-85.

Chapter Six   Modified Wool Fiber/ 改性羊毛纤维

1   Introduction/ 前言

Wool fibers **entangle** with each other under the combined action of moisture, thermal and mechanical forces due to **directional frictional effect** of **scale**s on wool fiber surface, natural crimp and high elastic recovery, which leads to volume reduction of fiber assembly. The phenomenon is called wool **felting**. Wool fabrics therefore show certain disadvantages such as shrinkage, no machine washability and **prickle**.

In recent years, with the improvement of life quality and consumption level, clothing is required not only to meet the needs of covering the body and keeping warm or cool, but also to be beautiful, light, comfortable, machine **washable**, **washable** and **wearable**, and easy **maintenance**. There is an irreversible tendency to develop high-count, light and thin wool fabrics. Since mid 80s, the grams per square meter of wool apparel fabrics for men and women have been decreased by 20%~30% and 10%~15%, respectively. According to foreign sample analysis, the high-count, light and thin wool fabrics (above 60^s) account for 40%~50% of all the worsted fabrics. The grams per square meter have been reduced by 41.6~62.5 g/m^2 compared with the traditional products of the same kind.

Various related regulations have been imposed at home and abroad with the penetration of high-**count**, light and thin wool products into people's live. At the 69th International Wool Spinning Conference, the regulation

由于羊毛纤维表面鳞片的定向摩擦效应、天然卷曲以及高度的弹性回复，在湿、热和机械外力的共同作用下，纤维间彼此纠缠，羊毛纤维集合体体积缩小，此现象称为羊毛缩绒。因此，羊毛产品具有缩水、不可机洗和刺痒感等缺点。

近年来，随着人们生活质量和消费水平的提高，服装不仅是遮体保暖的需求，而且要求美观、轻便、舒适、可机洗、洗可穿和易保养等。毛纺产品的高支、轻薄化已成为不可逆转的发展趋势。自20世纪80年代中期至今，男装和女装面料的平方米克重分别降低了20%~30%和10%~15%。根据国外来样分析，精纺呢绒面料中60支以上高支轻薄产品占40%~50%，与同类传统产品比较，平方米克重减少了41.6~62.5 g/m^2。

随着高支轻薄羊毛产品对人们生活的渗透，国内外已逐步出现了各种相关的标准。在第69届国际毛纺会议上，对有关

on the high-grade fabrics has been made. The wool fiber of Super 80^s~Super 200^s wool products should be within 19.5 μm (Super 80^s)~13.54 μm (Super 120^s) in diameter.

Furthermore, a **premium** price is paid for high-count wool. As is shown, wool fiber price and wool products added value increase with the decrease in diameter. The price rises more sharply for every reduction of 1 μm in diameter particularly when the diameter is less than 20 μm.

高档面料作了统一规定:"超级80支"到"超级200支"羊毛产品的纤维直径必须为19.5 μm(超级80支)到13.5 μm(超级200支)。

另外,高支羊毛的价格较昂贵。羊毛纤维直径和价格关系如图6-1所示。随直径减小,羊毛纤维价格和羊毛产品附加值增加。尤其当纤维直径在20 μm以下时,直径每减少1 μm,价格上升幅度更大。

2 Surface Modified Wool Fiber/ 表面改性羊毛纤维

Surface modified wool fiber with **shrinkproof** and machine washablility is obtained from eliminating the directional frictional effect caused by the scales on fiber surface, which is also known as mercerized wool fiber or shrinkproof wool fiber. Mercerized wool products are more lustrous because more scales are removed on mercerized wool fiber surface than those on shrinkproof wool fiber surface. They thus have a **silky** luster and cashmere-like feel. Therefore, the mercerized wool is praised as "cashmere-like wool".

Wool fiber by far can be subjected to surface modification including **strip**ping scale substraction treatment (degradation) and **resin** addition treatment (polymer **deposition**). The substraction treatment is to soften and remove fiber scales completely or partially by means of oxidants, enzymes or **plasma**, which may very well reduce the directional frictional effect. The addition treatment is to cover fiber surface with resin and other polymers to generate crosslinking reactions, so that the directional frictional effect could be lowered.

消除因纤维表面鳞片引起的定向摩擦效应,得到具有防缩和机可洗表面改性羊毛,也称丝光羊毛和防缩羊毛。丝光羊毛表面鳞片比防缩羊毛的消除得更多,丝光羊毛产品的光泽更亮,有蚕丝般光泽,有羊绒般手感,因此,丝光羊毛被誉为"仿羊绒羊毛"。

目前,羊毛表面改性包括剥鳞片减法处理(降解法)和树脂加法处理(高聚物沉积法)。减法处理是利用氧化剂、酶或等离子体等全部或部分软化和去除羊毛纤维表面的鳞片,减少羊毛表面的定向摩擦效应;加法处理是将树脂等高聚物覆盖在羊毛纤维表面以发生交联反应,使纤维定向摩擦效应减弱。

2.1 Oxidation Treatment/ 氧化法

Mechanism of wool fiber surface oxidation is shown in Figure 6–1. As is shown, the disulfide and peptide linkages in the scale cells are degradated by oxidants, leading to the damage of wool fiber scales. Primary oxidants in use include hypochlorite, **sodium dichloroisocyanurate** (DCCA), **potassium permanganate** (KMnO$_4$), sodium persulfate (Na$_2$S$_2$O$_8$), and H$_2$O$_2$ and so on.

羊毛纤维表面氧化机理如图 6-1 所示。利用氧化剂破坏鳞片细胞中二硫键和蛋白肽键等，从而破坏羊毛纤维鳞片。氧化剂主要有次氯酸盐、二氯异氰脲酸钠、高锰酸钾、过硫酸钠、双氧水等。

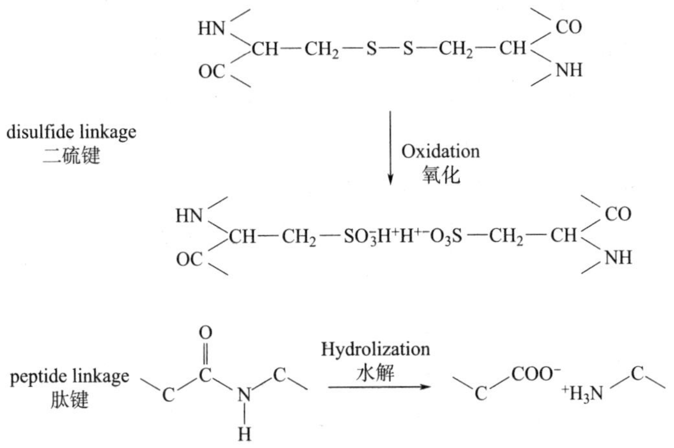

Figure 6–1　Mechanism of wool fiber surface oxidation
图6-1　羊毛纤维表面氧化机理

2.1.1　Wet Chlorination Treatment/ 湿氯化法

Wet **chlorination** treatment is to treat wool fiber in oxidant hypochlorite solution. The available **chlorine** form in the solution varies with the pH values, leading to the different reaction rates with wool fiber. Chlorination is usually sorted into acidic and alkaline ones according to the diverse pH values.

When wool fiber is placed in acid solution, especially in strong acid, it is difficult to obtain **even** treatment effect due to the fast reaction between free chlorine and wool fiber, so wool fiber is susceptible to be damaged. In this case, it is often recommended that chlorine **inhibitor** should be

湿氯化法是在氧化剂次氯酸盐溶液中处理羊毛的方法。溶液中有效氯的形式随 pH 不同而不同，不同形式的有效氯和羊毛反应的速率差异很大。根据氯化溶液 pH 不同，通常分为酸性氯化和碱性氯化。

在酸性条件下，尤其是强酸性时，由于自由氯和羊毛反应速度很快，很难获得均匀的处理效果，羊毛损伤也大。在溶液中加入抑氯剂，减缓氯化速率，提高

added into the solution to slow down the chlorination rate and improve the treatment uniformity. DCCA and its salts are the most extensively used as acid **chlorinat**ing agents. When pH value ranges from 3 to 5, the chlorine is gradually released into the solution. The chlorination reaction therefore is mild and even, leading to small damage or slight yellowing to wool fiber. Commercially available **anti-felting** agents containing DCCA are Basolan (BASF) DC, ACL-60 (Monsanto) and CDS 60 (FMC Corp), etc.

Wool fiber reacts mildly with hypochlorous acid (HClO) or ClO$^-$ in alkaline solution and it is easy to get even treatment. However, HClO or ClO$^-$ gradually penetrates into the **cortex** with increasing time, resulting in greater damage to wool fiber. In addition, the treated wool fiber is often degraded by alkali, becomes yellow and feels rough. KMnO$_4$ or peroxymonosulfuric acid therefore is generally employed in alkaline or **neutral** chlorination. Commercially available products for alkaline chlorination are Dylan ZB (Stevenson's) and Harriset (Milton Harris Associates), etc.

处理的均匀性。广泛应用的酸性氯化剂是二氯异氰脲酸及其盐，在pH=3~5时，在溶液中逐渐释放出氯，氯化过程温和、均匀，对羊毛损伤小，泛黄程度也较小。用DCCA作为防毡缩剂的商品有Basolan（BASF）DC、ACL-60（Monsanto）和CDS 60（FMC Corp.）等。

在碱性条件下，HClO或ClO$^-$和羊毛的反应速度慢，容易获得均匀的处理效果，但随着时间的延长，它们逐渐渗透到皮质层，羊毛损伤较大。另外，碱会促使羊毛降解，引起泛黄，使手感粗糙。因此，在碱性或中性氯化中，通常加入高锰酸钾或过一硫酸等。用于碱性氯化的商品有Dylan ZB（Stevenson's）和Harriset（Milton Harris Associate）等。

2.1.2　Dry Chlorination Treatment/ 干氯化法

Dry chlorination treatment has been started by England Wool Industry Research Associate (WIRA) and Australia Confederation Science and Industry Research Organization (CSIRO). The former is to treat wool fiber (moisture content is 5%~7%) with chlorine in a vacuum. The obtained wool fiber achieves excellent and even anti-felting effect without any damage, yet it needs to be softened due to the stiff handle. And the costs of equipment and the processing are relatively high. The latter is to treat wool fiber (moisture content is 9%) with a mixture of air and chlorine at atmosphere pressure. The obtained wool fiber feels very soft. And the costs of equipment and processing are relatively

英国羊毛工业研究会（WIRA）和澳大利亚联邦科学与工业研究组织（CSIRO）开发了干氯化法工艺。前者是在真空中，使用氯气处理含湿量为5%~7%的羊毛，羊毛可获得极佳而均匀的防毡缩效果，且羊毛无损伤，但手感粗糙，还要经过柔软处理，另外，设备和加工成本较高。后者是在常压下，使用空气和氯的混合气体处理含湿量为9%的羊毛，羊毛手感柔软，

low. This treatment therefore has been extensively used in Australia, New Zealand and other countries.

设备和加工成本较低，在澳大利亚和新西兰等国家被广泛应用。

2.1.3 KMnO$_4$ Treatment/ 高锰酸钾法

KMnO$_4$ was comprehensively adopted to treat wool fiber in 60s. Wool fiber should be **agitate**d in **saturated** salt solution otherwise the shrinkproof effect is not obvious. Besides, the excessive salt consumption may result in salting out and equipment **corrosion**. At the same time, the colored **manganese** particles deposited on the fiber surface are not easy to remove.

20世纪60年代广泛使用高锰酸钾处理羊毛，但是需在饱和的盐溶液中使用，否则防缩效果不明显。此外，盐用量过大，会产生盐析和设备腐蚀，同时，沉积在纤维表面的有色锰粒子不容易去除。

2.1.4 Persulfate Treatment/ 过一硫酸法

The treatment of wool fiber with persulfate has mild and **uniform** effects, and no yellowing. There is no obvious anti-felting though. Thus other shrinkproof methods have to be employed to meet machine washable requirement. Furthermore, Wool fiber is unstable to hard water and the generation of **calcium sulfate** (CaSO$_4$) makes its feel rough.

过一硫酸处理羊毛，作用缓和，效果均匀，不泛黄，但是防缩效果不明显，还需要采用其他的防缩方法以达到机可洗，另外，羊毛纤维对硬水不稳定，会产生硫酸钙，使其手感粗糙。

2.2 Enzyme Treatment/ 酶处理

2.2.1 Proteinase Classification/ 蛋白酶分类

Wool fiber belongs to protein fiber and it is often recommended that **proteinase** be applied to decompose the protein peptide bonds in wool fiber. According to pH value of **catalytic** hydrolysis reaction, proteinase is divided into acid (pH=2.5~5.0) proteinase, neutral (pH=7.0~8.0) proteinase and alkaline (pH=9.5~10.5) proteinase. According to origin, proteinase is sorted into animal proteinase, plant proteinase and microbe proteinase. According to the way of protein hydrolysis, proteinase is classified into **endonuclease** and **exonuclease**. For endonuclease, peptide bonds inside the

羊毛属于蛋白质纤维，因此，使用蛋白酶催化分解羊毛中的蛋白肽键。按催化水解反应的pH，蛋白酶分为酸性蛋白酶（pH=2.5~5.0）、中性蛋白酶（pH=7.0~8.0）和碱性蛋白酶（pH=9.5~10.5）；按来源，蛋白酶分为动物蛋白酶、植物蛋白酶和微生物蛋白酶；按蛋白质水解的方式，蛋白酶分为内切

protein molecular are split to create polypeptide with low molecular weight. For exonuclease, terminal peptide bonds of protein or polypeptide molecule amino or carboxyl are split to liberate amino acids.

酶（切开蛋白质分子的内部肽键生成分子量较小的多肽类）和外切酶（切开蛋白质或多肽分子氨基或羧基末端的肽键而游离出氨基酸）。

2.2.2　Catalytic Process/ 催化过程

Enzyme is the catalyst for the peptide bond hydrolysis during the reaction of enzyme and wool fiber. In proper condition, enzyme **diffuse**s into the fiber interior from the hydrophilic part of the scales on the wool fiber surface, leading to peptide bond hydrolysis, so that the protein partly dissolves. The steps of catalytic process are: (1) Enzyme diffuses in solution to the fiber surface. (2) Enzyme is adsorbed on fiber surface. (3) Enzyme diffuses into the fiber interior. (4) The **hydrolyz**ing reaction occurs by enzyme **catalysis**.

在酶与羊毛纤维的反应体系中，酶是肽键水解的催化剂。在适宜的条件下，酶从羊毛纤维表面鳞片层的亲水部分进入纤维内部，促使羊毛中的肽键水解，使部分蛋白质溶解。其催化过程为：（1）酶在溶液中向纤维表面扩散；（2）酶被纤维表面吸附；（3）酶向纤维内部扩散；（4）酶催化水解反应。

2.2.3　Characteristics/ 特点

Enzyme treatment exhibits many desirable characteristics such as low energy consumption, high efficiency and little **contamination** to environment. It also improves anti-felting, softness and moisture absorption of wool fiber. The treated wool fiber is considered to be cashmere-like in handle instead of **itchy** sensation. However, it is hard to accurately control enzyme performance and treatment process.

酶处理具有能耗低、效率高和污染小等特点，改善羊毛的防毡缩、柔软和吸湿等性能，处理过的羊毛没有刺痒感，具有山羊绒的手感。但酶性能和处理过程较难准确控制。

2.3　Plasma Treatment/ 等离子体处理

2.3.1　Plasma/ 等离子体

Plasma is known as the fourth state of matter. It is an aggregation of **charged positive ions**, **negative ion**s, electrons, **free radical**s and various radicals. Plasma is an

等离子体被称为物质的第四态，由带电的正离子、负离子、电子、自由基和各种活性基团等

ionized gas preserving electronic **quasi-neutrality**, which is generated by exposing gas or gas mixture (air, oxygen, nitrogen, etc.) on high energy condition such as electric field, high temperature, **laser beam**, **radiation** and so on.

Plasma is divided into high temperature, thermal and cold ones according to the temperature. Cold plasma so far has been extensively used in fiber surface modification due to its high energy and low temperature. For one thing, the electrons have sufficient energy to excite, **dissociate** and ionize **reactant**s. For another, the reaction system keeps low temperature, which reduces energy consumption and saves cost.

组成的集合体。等离子体是呈电中性的电离态气体，由空气、氧气和氦气等气体或气体混合物在电场、高温、激光束和辐射等高能量条件下发生电离而产生。

根据温度，等离子体分为高温等离子体、热等离子体和冷等离子体。冷等离子体具有较高能量和较低温度，一方面电子具有足够的能量使反应物分子激发、离解和电离；另一方面反应体系保持低温，减少能耗和节约成本。因此，冷等离子体目前广泛应用在纤维表面改性中。

2.3.2 Plasma Treatment Mechanism/ 等离子体处理机理

Active species in plasma physically or chemically **etch** the scales on the wool fiber surface, resulting in morphological and chemical changes as well as properties modification. Specific surface area and surface roughness increase while directional friction coefficient reduces due to the scale damage. Besides, the introduction of oxygen-containing groups supplements the polar groups so that the hydrophilicity increases. All these contribute to the improvement of anti-felting of wool fiber.

等离子体中的活性粒子对羊毛纤维表面进行物理刻蚀和化学刻蚀，从而使其形态结构、化学结构和性能发生变化。一方面羊毛纤维表面鳞片被破坏，比表面积和表面粗糙度增加，定向摩擦系数减小；另一方面引入含氧基团，极性基团增加，亲水性提高，这些都有助于羊毛缩绒性的改善。

2.3.3 Change in Morphology/ 形态变化

Figure 6-2 shows the SEM images of wool fiber before and after plasma treatment. As shown in Figure 6-2(a) and Figure 6-2(b), wool fiber has a relatively smooth surface, **overlap**ping scales and distinct edge before treatment. After that, small particles appear on the fiber surface and cracks emerge on the scale edge partially. Some of scales even **peel off** the fiber surface. Plasma physically etching effect on

等离子体处理前后羊毛纤维扫描电子显微镜照片如图 6-2 所示。处理前，羊毛纤维表面相对光滑，鳞片完整，边缘清晰。处理后，羊毛纤维表面出现小颗粒物，鳞片边缘出现裂痕，部分鳞片脱落。等离子体对羊毛纤维表

the fiber surface increases with the treatment time. Oxygen plasma has the strongest **destructive** effect on the wool fiber surface, followed by **argon** plasma, and **nitrogen** plasma is the weakest.

面的物理刻蚀作用随着处理时间的增加而增强。氧等离子体对羊毛纤维表面的破坏作用最强，其次是氩等离子体，氮等离子体最弱。

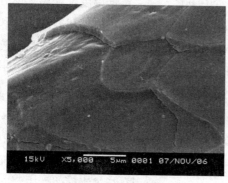
(a) Before treatment/处理前

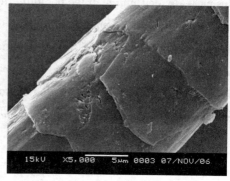
(b) After treatment/处理后

Figure 6-2　Longitudinal morphology of wool fiber before and after plasma treatment
图6-2　等离子体处理前后羊毛纤维纵向形态

2.3.4　Changes in Chemical Composition/ 化学成分变化

Table 6-1 and Table 6-2 show the chemical compositions and chemical group contents of wool fiber surface before and after plasma treatment, respectively. After plasma treatment, carbon content decreases while oxygen and nitrogen contents increase. The content of C—C decreases and the contents of other nitrogen-containing and oxygen-containing polar groups increase.

等离子体处理前后羊毛纤维表面化学成分和化学基团含量分别见表6-1和表6-2。处理后，碳含量减少，氧和氮含量增加，C—C减少，其他含氮和含氧等极性基团增加。

Table 6-1　Chemical compositions and atomic ratios of wool fiber surface before and after plasma treatment
表6-1　等离子体处理前后羊毛纤维表面化学成分及其比例

Sample 试样	Chemical composition 化学成分（%）				Atomic ratio 原子比例	
	C 碳	O 氧	N 氮	S 硫	O/C 氧/碳	N/C 氮/碳
Before treatment 处理前	79.5	10.7	7.3	2.5	0.135	0.091
After treatment 处理后	70.4	16.2	10.8	2.6	0.209	0.049

Table 6-2 Chemical group contents of wool fiber surface before and after plasma treatment
表 6-2 等离子体处理前后羊毛纤维表面化学基团含量

Sample 试样	Chemical group content 化学基团含量（%）			
	C—C 碳—碳	C—O/C—N 碳—氧/碳—氮	—COO— 羧基	O=C—N 酰胺基
Before treatment 处理前	73.2	18.1	6.6	2.1
After treatment 处理后	48.9	29.6	13.9	7.6

2.3.5 Characteristic/ 特点

Plasma treatment is a **nonaqueous** and environmental-friendly process due to the absence of water and chemicals. The plasma treatment involves only a thin layer on the substrate surface in depth. Physical or chemical changes typically occur at the surface of 50~100 nm in depth. Fiber tenacity, feel and other properties therefore are not affected. Plasma treatment has many desired characteristics including no pollution to the environment, low damage to the fiber, short processing time and high efficiency.

由于不使用水和化学试剂，等离子体处理是一个干燥和环境友好的过程。等离子体处理仅涉及基体表面极薄的一层，一般在离基体表面 50~100 nm 的表层发生物理或化学变化，纤维的强度和手感等性能不受影响。因此，等离子体处理具有无环境污染、纤维损伤小、处理时间短和效率高等优点。

2.4 Polymer Treatment/ 聚合物处理

Polymer treatment is to make resin form polymer on the surface of wool fiber, which reduces the directional friction effect (DFE) and improves anti-felting of wool fiber. It could be explained by followings. (1) Resins are crosslinked among fibers by "spot welding" so that the fibers may be fixed and DFE may not work. This mechanism is considered suitable for the small amount of resin. (2) The film on the fiber surface is formed by resin, which could shade or cover the scales completely. (3) A large amount of resin is deposited on the surface of the fibers, which prevents the interaction between the scales.

聚合物处理是利用树脂在羊毛纤维表面形成高分子聚合物，降低定向摩擦效应，改善羊毛纤维防毡缩性。其防毡缩原理主要有三种理论：（1）树脂在纤维与纤维之间通过"点焊接"产生交联，从而使纤维固定化而不能移动，定向摩擦效应不起作用，这种机理适用于树脂量较少的情况；（2）树脂在纤维表面形成薄膜，鳞片被遮蔽或者完全包覆；

（3）大量树脂沉积在纤维表面，阻止了鳞片表面之间相互作用。

3 Slenderized Wool Fiber/ 细化羊毛纤维

3.1 Decrement Slenderization (Mercerization) Treatment/ 减量细化（丝光）法

Decrement slenderization treatment was originally employed to be anti-felting, in which some or whole scales on wool fiber surface are **corrosively** removed to achieve mercerizing and slenderizing effect. Essentially disulfide bonds in **keratin** polymer are chemically or biochemically destroyed. The wool surface is consequently **erode**d and the fiber gradually becomes finer. Decrement treatment enhances fiber luster but the slenderizing effect is not obvious. Wool fiber is only decreased by 1.5~2 μm in diameter. However, the scales on wool fiber surface are severely damaged, resulting in low material utilization. Taking the wool fiber of 20 μm in diameter for instance, if the diameter is reduced by 1 μm, the wool decrement reaches about 10%. It is concluded that the raw material suffers a great loss by this treatment.

减量细化法的最初目的是防毡缩，其基本原理是：将羊毛表面鳞片部分或全部腐蚀掉，以实现丝光和变细的效果。实质就是采用化学或生物化学的方法，破坏角质中的二硫键结构，将羊毛表层剥蚀，客观上使得纤维变细。减量处理可使纤维光泽变亮，但细化效果不明显，羊毛纤维的直径仅下降1.5~2 μm。羊毛纤维表面的鳞片被严重损伤，导致原材料的利用率大大降低。以直径为20 μm 的羊毛纤维为例，如果纤维直径减少1 μm，羊毛的减量约为10%。因此，原材料损失较大。

3.2 Stretching Slenderization Technology (SST)/ 拉伸细化技术

3.2.1 Mechanism/ 机理

The main chemical composition in wool fiber is keratin, which exists in two forms. One is **helical** chain, i.e. α-keratin, and the other is linear **folding** chain, i.e. β-keratin. The keratin in wool fiber generally appears as α-**helix** that will convert into β **configuration** when a longitudinal force acts on wool fiber. This **transformation** of spatial **conformation** results in the stretching and

羊毛纤维的主要化学成分是角朊，角朊有两种存在形式，一种是螺旋链，即α型角朊；另一种是直线状的折叠链，即β型角朊，羊毛纤维的角朊通常呈α型的螺旋链，当纵向力作用于羊毛纤维上时，大分子链由α型转变

slenderization of wool fiber.

成 β 型。这种空间构象的转变使羊毛纤维伸长和变细。

3.2.2 Preparation Process/ 制备工艺

At the very beginning, chemical pretreatment is employed to destroy the disulfide bonds, salt bonds and hydrogen bonds among fiber polymers, which is beneficial for the transformation of α-helix into β-keratin. After that, the pretreated wool fiber is drawn by physical treatment. Finally, the stretched wool fiber is exposed to hot and humid condition to be **permanently** set. Slenderized wool fiber thereby is achieved.

首先，采用化学预处理破坏羊毛纤维大分子链之间的二硫键、盐式键和氢键，有利于羊毛由 α 型转变到 β 型。其次，采用物理方法拉伸经过预处理的羊毛纤维，最后，在湿热条件下永久定型，从而获得细化羊毛。

3.2.3 Changes in Molecular Structure/ 分子结构变化

Wool molecular chain **transform**s from α-helix to β-**pleated** sheet structure after slenderization. The orientation and crystallinity in slenderized wool fiber are correspondingly increased. It is estimated that the crystallinity of slenderized wool fiber is 3 times that of common wool fiber. Significant change in the **secondary structure** of wool fiber chiefly happens during the initial period of slenderization. There is no significant change in the structure of slenderized wool fiber after that.

细化后，羊毛大分子链从 α 螺旋转变为 β 折叠结构。细化羊毛纤维的取向度和结晶度增加，细化羊毛纤维的结晶度是羊毛纤维的3倍。细化初期，羊毛纤维次生结构发生明显变化，然后，细化羊毛纤维结构的变化不显著。

3.2.4 Changes in Morphology/ 形态结构变化

The cross section and longitudinal form of wool fibers before and after slenderization are shown in Figure 6-3 and Figure 6-4, respectively. Slenderized wool fiber is remarkably finer in diameter compared with common wool fibers. The cross section has changed from circular to non-circular. Slenderized wool fiber has similar scale structure to cashmere, which is relatively finer and longer, with lower density and larger distance between scales. Besides, slenderized wool fiber surface is relatively smoother.

细化前后羊毛纤维横截面和纵向形态分别如图6-3和图6-4所示。细化羊毛纤维的横截面由圆形变为非圆形，其直径相比一般羊毛纤维明显变小。细化羊毛纤维的鳞片结构类似于羊绒，鳞片细长，密度减小，鳞片间距离增大。此外，细化羊毛纤维的表面较光滑。

 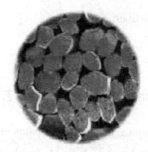

(a) Before SST/细化前　　　　　　(b) After SST/细化后

Figure 6–3　Cross sectional morphology of wool fiber before and after SST
图6-3　细化前后羊毛纤维横截面形态

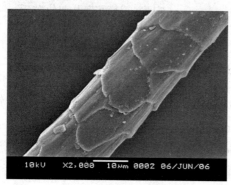 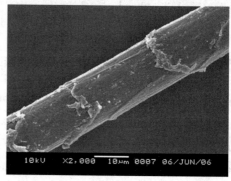

(a) Before SST/细化前　　　　　　(b) After SST/细化后

Figure 6–4　Longitudinal morphology of wool fiber before and after SST
图6-4　细化前后羊毛纤维纵向形态

3.2.5　Changes in Properties/ 性能变化

Fineness and length: Fineness and length of protein fibers are shown in Table 6–3. It indicates that the fineness and length of slenderized wool fiber are decreased and increased compared with common wool fiber, respectively. Fineness CV and length CV are increased as well. Slenderized wool fiber is similar to cashmere in fineness. And it is about 1.2~1.6 times longer than common wool fiber. At 110% elongation, slenderized wool fiber is reduced by 6~7 μm in diameter.

Frictional coefficient: The static frictional coefficient and effect of protein fibers are shown in Table 6–4. As the

细度和长度：蛋白质纤维的细度和长度见表 6-3。与一般羊毛纤维相比，细化羊毛纤维细度减小，长度增加，细度和长度的不匀率增加。细化羊毛的细度与羊绒基本相同，细化羊毛的长度是普通羊毛的 1.2~1.6 倍。110% 的拉伸使细化羊毛纤维直径减少 6~7 μm。

摩擦系数：蛋白质纤维的静摩擦系数及效应见表6-4。由

Table 6-3 Fineness and length of protein fibers
表6-3 蛋白质纤维的细度和长度

Fiber 纤维	Fineness 细度（mm）	Fineness CV 细度不均率（%）	Length 长度（mm）	Length CV 长度不均率（%）
Cashmere 羊绒	15.38	20.2	—	—
Wool 羊毛	18.50	21.2	86.8	33
Slenderized wool 细化羊毛	16.20	23.5	119.4	40

Table 6-4 Static frictional coefficient and effect of protein fibers
表6-4 蛋白质纤维的静摩擦系数及效应

Fiber 纤维	Static frictional coefficient 静摩擦系数		Frictional effect 摩擦效应（%）
	Frontward 顺向	Backward 逆向	
Wool 羊毛	0.1557	0.1726	5.15
Slenderized wool 细化羊毛	0.2600	0.2757	2.9

angle and density of the scales decrease, the static frictional coefficient of wool fiber is enhanced after slenderization, and the directional frictional effect is declined. Slenderized wool fiber therefore has a cashmere-like feel and excellent anti-felting.

Dyeability: Slenderized wool fiber has higher dyeing rate than superfine and other fine wool fibers. It ranks first in dyeing rate at low temperature.

Tensile property: Slenderized wool fiber has relatively lower breaking strength, elongation at break and **fracture work**, but higher breaking tenacity compared with common wool fiber.

于鳞片倾斜角和密度减小，细化后，羊毛纤维的静摩擦系数增加，定向摩擦效应减小，因此，细化羊毛具有羊绒般的手感和较好的防毡缩性。

染色性： 细化羊毛纤维的上染速率较高，相比超细羊毛纤维和其他细支羊毛纤维而言，细化羊毛纤维的低温上染速率最高。

强伸性： 细化羊毛纤维的断裂强力、断裂伸长和断裂功低于羊毛纤维，而断裂强度高于普通羊毛纤维。

3.2.6 Slenderized Wool Products/ 细化羊毛产品

Slenderized wool fibers developed by SST in Japan

日本和澳大利亚开发的细

and Austria are registered trademarks—Odin and Optim, respectively.

Permanent-setting slenderized wool (OptimFine): After slenderization, the cross section changes from near circular to polygon. The fiber diameter is reduced by 3~4 μm, and becomes finer, stronger and more lustrous. Optim fine has similar structures and properties as silk and superfine wool fiber. It has β-pleated sheet structure as silk. Light fabric made by Optim fine has silk-like feel, excellent drapability and elegant luster.

Shrinkable-expansion slenderized wool (OptimMax): It is a shrinkable slenderized wool fiber that shrinks by 20%~25% in length. The bulkiness may be increased by 20% when it is blended spun with wool fiber. This particular yarn is suitable to be knitted to form a **bulky**, light weight fabric. The fabric weight is reduced and the fluffy degree is increased at the same time, which has opened up the market for the casual wool textiles and met the requirements of the light weight fabric.

永久定型细化羊毛（OptimFine）：细化后，纤维横截面由近圆形变成多角形，直径减小3~4 μm，更细更强，光泽更亮。Optim fine 羊毛的结构和性能接近于丝和超细羊毛纤维，具有丝纤维中的β折叠结构，织造出的轻薄织物具有丝质手感、优良的悬垂性和优雅的光泽。

可回缩膨胀细化羊毛（OptimMax）：是一种可收缩的细化羊毛，长度可收缩20%~25%。与羊毛混纺的纱线的蓬松度增加20%，这种纱线可以制作成蓬松轻薄型的针织物，在减轻织物重量的同时增加了蓬松度，为休闲羊毛纺织品创造市场，并适应市场对轻薄织物的要求。

4　Bulking Wool Fiber/ 膨化羊毛纤维

Natural crimp significantly influences spinnability and handle of wool fiber. Bulking wool fiber is obtained by altering the crimp form to improve the spinnability of wool fiber, yarn quality and wool product performance.

Bulking wool fiber was first developed by wool research organization in New Zealand. Wool top is stretched, heated (non-permanent set), and **contract**ed after **relaxation** in order to obtain bulking wool fiber. When made into the clothes with the same **specification**, it is estimated that bulking wool fiber could be saved about 20% of wool fiber,

天然卷曲显著影响羊毛纤维的可纺性和手感。通过改变羊毛纤维卷曲形态，提高羊毛可纺性，改善成纱品质和羊毛产品的性能，得到的羊毛被称为膨化羊毛。

新西兰羊毛研究组织最早开发了膨化羊毛。羊毛条经拉伸、加热（非永久定型），松弛后收缩以得到膨化羊毛纤维。膨化羊毛制成同等规格的服装，据估计可节省羊毛约20%，且具有更

but with better heat retention, bulkier, softer and more comfortable handle. Bulking wool fiber is blended spun with wool fiber to develop bulky or super bulky yarns and knitted fabrics, which provides conditions for the development of light weight, casual and sports wool products.

Chemical treatment: When wool fiber is placed in ammonia, ammonia molecules penetrate into the interior of wool fiber with **bilateral** structure, which causes wool fiber to excessively shrink, resulting in the formation of crimp. The crimp of wool fiber is **stabilize**d through heat setting.

Super-crimping treatment: After the wool top with bilateral structured wool fibers is drafted by the **roller**s, it is **relax**ed to form crimp in a free state. The crimp is stabilized by heat setting in stream. Stretching damages partial disulfide bonds and salt bonds. Different strains and **stress**es of **orthocortex** and cortex lead to various relaxing resilience. All these contribute to more three-dimensional crimps. Properties of wool fibers before and after super-crimping treatment are shown in Table 6-5.

化学处理：羊毛纤维浸泡氨水中，氨分子渗入具有双侧结构的羊毛纤维内部，引起羊毛纤维过度收缩而产生卷曲，再经过热定型使卷曲稳定。

超卷曲加工：含双侧结构毛纤维的毛条经过罗拉牵伸后，在自由状态下松弛而形成卷曲，然后在蒸汽中定型使卷曲稳定。拉伸使二硫键和盐式键部分断裂，正偏皮质层的应变和应力不同，松弛时的回复也不同，从而产生更多的三维卷曲。超卷曲处理前后羊毛纤维的性能见表6-5。

Table 6-5 Properties of wool fibers before and after super-crimping treatment
表6-5 超卷曲处理前后羊毛纤维的性能

Index 指标	Before treatment 处理前	After treatment 处理后	Changing ratio 变化率（％）
Crimp number 卷曲数（个/cm）	2.16	2.62	21.3
Crimp ratio 卷曲率（％）	3.73	8.54	129.0
Residual crimp ratio 剩余卷曲率（％）	2.89	7.21	149.5
Crimp elasticity 卷曲弹性率（％）	77.68	84.89	9.3
Breaking tenacity 断裂强度（cN/dtex）	1.87	2.06	10.2
Breaking elongation ratio 断裂伸长率（％）	43.44	40.40	-7.0
Initial modulus 初始模量（cN/dtex）	29.2	34.6	18.5

Twisting treatment: The wet top is stretched after being twisted in the air at 20 ℃ and relaxed in no time to generate crimp.

加捻法：湿润的毛条加捻后在20 ℃的空气中拉伸，并立即松弛而产生卷曲。

5 Colored Wool Fiber/ 彩色羊毛纤维

Blue sheep have been successfully bred in Australia already. The color contains light blue, sky blue and ocean blue. During years of research, Russian experts in febrile animal husbandry have found that different trace metal elements fed to the sheep change the wool color. For example, the iron and copper elements change the wool color into light red and light blue, respectively. Colored sheep containing light red, light blue, golden yellow and light grey have been bred so far. The color of colored wool textiles is still bright as before and does not fade after being blown, sunbathed, rained and washed. Colored wool fiber does not need to be dyed, nor is any chemistry involved. It is not corroded. Therefore, it has high tenacity and good resilience. The clothes made of colored wool fiber are resistant to abrasion and wear, and have long service life.

澳大利亚已成功培育出蓝色羊，颜色包括浅蓝、天蓝到海蓝。俄罗斯的畜牧专家经多年研究发现，给羊喂不同的微量金属元素，能改变羊毛的颜色。例如，铁元素和铜元素可使羊毛变成浅红色和浅蓝色，目前，已成功培育出浅红色、浅蓝色、金黄色及浅灰色等彩色羊。彩色羊毛纺织品经风吹、日晒、雨淋和洗涤，颜色仍然鲜艳如初，毫不褪色。彩色羊毛因不需染色，不含化学物质，未被腐蚀，因此，强度高、弹性好，制成的服装耐磨、耐穿、使用寿命长。

Exercises/ 练习

1. Why wool fiber needs to be modified?
2. Describe the surface modification methods of wool fibers and their mechanism.
3. Elaborate the wool stretching mechanism and its changes in structure and property.

References/ 参考文献

[1] WANG Chunxia, QIU Yiping. Two sided modification of wool fabrics by atmospheric pressure plasma jet: influence of processing parameters on plasma penetration[J]. Surface & Coatings Technology, 2007, 201(14) : 6273–6277.

［2］王春霞，邱夷平. 氦气/氧气常压等离子射流处理对羊毛织物性能的影响［J］. 毛纺科技，2008（1）：1-3.

［3］卢可盛. 低温常压等离子体处理羊毛改性技术研究［J］. 上海纺织科技，2004，32（2）：31-33+59.

［4］唐琴. 低温等离子体处理羊毛织物改性研究［J］. 毛纺科技，2009，37（9）：22-24.

［5］金郡潮，戴瑾瑾. 等离子体处理羊毛织物防毡缩性能的研究［J］. 印染，2002（1）：31-32.

［6］朱若英，滑钧凯，黄故，等. 羊毛低温等离子体处理后的染色性能研究［J］. 天津工业大学学报，2002，21（4）：23-24.

［7］朱宝瑜，张一心，王维. 拉伸改性羊毛特性测试与比较研究［J］. 毛纺科技，2004（1）：5-8.

［8］张淑洁. 拉伸细化羊毛纤维的性能及其产品开发［D］. 天津：天津工业大学，2003.

［9］周勤谦. 拉伸羊毛的性能及膨化羊毛的纺纱技术［J］. 毛纺科技，2004（3）：5-8.

［10］徐婕，于鹏美，关晋平. 氧化还原法表面改性羊毛的理化性能［J］. 纺织学报，2014，35（7）：1-7.

Part Three

Synthetic Fiber/ 合成纤维

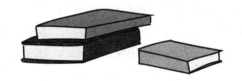

Chapter Seven Polytrimethylene Terephthalate (PTT) Fiber/ 聚对苯二甲酸丙二醇酯纤维

1 Introduction/ 前言

Research and development of new materials have become a strong backing for all countries in the world toward the 21st century. Synthetic fiber occupies very substantial place in textile industry. Polyester fiber ranks the first among synthetic fibers due to its excellent properties and low price. The long chain macromolecule of polyester fiber is composed of ester at least 85% by weight. According to the number of hydroxyl group, three members of the polyester **generic** group are commercially possible: **PET (polyethylene terephthalate)**, **PBT (polybutylene terephthalate)** and PTT (polytrimethylene terephthalate). Among them, PET and PBT fibers have already been in massive industrial production and extensive application. PTT fiber, which also belongs to the polyester generic group, exhibits the characteristics of both polyester and nylon such as good **antifouling**, easy dyeing, soft touch, elasticity and easy drying. It could tell that PTT will gradually replace PET and nylon in the near future, and become one of the most popular synthetic fibers in the 21st century.

In 1995, PTT was successfully formed by Shell Company in the United States with **ethylene** as raw material to produce 1, 3-**propanediol** (PDO). DuPont Company has developed a biological **enzymatic** process

新材料的研制和开发成为世界各国迈向21世纪的坚实后盾。合成纤维在纺织业中占有非常重要的地位。聚酯纤维以其优良的性能和低廉的价格成为第一个合成纤维品种，其中长链分子中酯含量不低于85%。根据羟基数目，商用聚酯有PET（聚对苯二甲酸乙二醇酯）、PBT（聚对苯二甲酸丁二醇酯）和PTT（聚对苯二甲酸丙二醇酯）三类。其中PET纤维和PBT纤维已有大量的工业化生产和广泛的应用。而同样作为聚酯纤维家族的PTT纤维兼有涤纶和锦纶的特性，除防污性能好外，还有易于染色、手感柔软、富有弹性、易干等特性。因此，在不久的将来，PTT将逐步替代涤纶和锦纶而成为21世纪最热门的合成纤维之一。

1995年，美国Shell公司以乙烯为原料生产1，3-丙二醇（PDO）成功开发了新型纺丝聚合物PTT。杜邦（DuPont）公

for the preparation of PDO. PDO is obtained by **microbial fermentation** of low-cost **carbohydrates** and then processed into PTT. The method is relatively simple and the cost is low. Asahi Kasei, Toray, Teijin and other Japanese companies successively have contracted with Shell and DuPont with attempt to jointly lead the development of PTT fiber and its products. Degussagn Company and Zimmer Company in Germany co-developed PTT fiber production process in certain economic scale. At the end of 2002, the price of PTT chips has been reduced by 25% due to the expansion of production scale, and will continue to decline in the future. In addition, with the extraordinary performance of PTT fiber, its comprehensive application is bound to come.

司开发出一种生物酶工艺制备 PDO，用低廉的碳水化合物通过微生物催化发酵的方法可以制得 PDO，其生产的 PDO 用于生产 PTT，方法简单、成本低。日本旭化成、东丽、帝人等公司先后与 Shell 和 DuPont 缔结合约企图引领 PTT 纤维及其制品的开发。德国的 Degussagn 和 Zimmer 公司又共同开发出一定经济规模的 PTT 纤维生产工艺。2002 年末，由于生产规模的扩大，PTT 切片已降价 25%，今后还会继续降低，加之 PTT 纤维的优异性能，决定了 PTT 纤维的应用高峰必将到来。

2 Manufacturing/ 制备

2.1 Preparation of 1, 3-propanediol (PDO)/1, 3- 丙二醇制备

PDO is the main **monomer** to prepare PTT. At present, the industrial production of PDO can be realized mainly by **ethylene oxide (EO)** method in Shell Company, **propionaldehyde** method in Degussa Company and boenzymatic catalysis method in DuPont Company.

PDO 是制备 PTT 的主要单体，目前，能实现 PDO 工业化生产的主要有 Shell 公司的环氧乙烷法、Degussa 公司的丙醛法和 DuPont 公司的生物酶催化法。

2.1.1 EO method/ 环氧乙烷法

PDO is prepared by EO method in Shell Company. EO method is a two-step process. The conversion ratio are 55% and 89%, respectively. The reactions are as follows: Ethylene reacts with oxygen to form EO. The obtained EO then reacts with CO and H_2 to produce PDO in the presence of catalyst. The reaction occurs at 105 ℃ for 3 hours and

Shell 公司采用环氧乙烷法制备 PDO。环氧乙烷法分为两步，转化率分别为 55% 和 89%。其反应如下：乙烯与氧反应生成环氧乙烷，环氧乙烷在催化剂作用下与 CO 和 H_2 反应制得 PDO。

the molar ratio of CO to H_2 is 1∶1. The key to EO method is the preparation and selection of catalysts. The equipment investment is high, but raw materials are easily available and cheap.

CO 与 H_2 的摩尔比为 1∶1，反应温度为 105℃，反应时间为 3 h。环氧乙烷法的关键在于催化剂的制备与选择。设备投资大，但原料易得、价格低。

$$H_2C=CH_2 + O_2 \xrightarrow{\text{Catalyst/催化剂}} H_2C\underset{O}{-}CH_2$$

$$H_2C\underset{O}{-}CH_2 + CO + H_2 \xrightarrow{\text{催化剂}} HOCH_2CH_2CH_2OH$$

2.1.2 Propionaldehyde Method/ 丙醛法

Propylene is adopted to prepare PDO in Degussa Company in Germany. The reactions are as follows: Firstly, **propylene** is **oxidated** into acraldehyde. Then, **3-hydroxyl propylaldehyde** (HPA) is formed by the double bond hydration of **acraldehyde** with water in the presence of catalyst. Finally, HPA reacts with H_2 to produce PDO under the action of Ni and other catalysts. The product yield depends on the **hydration** reaction of acraldehyde. The quality of the final product is determined by the **hydrogenation** effect of HPA. The key technology of the last two steps is the choice of catalyst.

德国 Degussa 公司采用丙烯制备 PDO，其反应如下：首先丙烯氧化成丙烯醛，然后丙烯醛在催化剂作用下与水进行双键水合形成 3-羟基丙醛（HPA），最后 HPA 在镍等催化剂作用下与 H_2 反应生成 PDO。产品的收率取决于丙烯醛水合反应，最终产品的质量则由 HPA 的加氢效果决定，后两步反应的关键技术是对催化剂的选择。

$$CH_2=CHCH_3 + O_2 \longrightarrow CH_2=CHCHO$$
$$CH_3CH_2CHO + H_2O \longrightarrow HOCH_2CH_2CHO$$
$$HOCH_2CH_2CHO + H_2 \longrightarrow HOCH_2CH_2CH_2OH$$

2.1.3 Enzyme Catalysis Method/ 酶催化法

PDO is prepared by biologic enzyme catalysis synthesis technology in DuPont Company in the United States. Carbohydrate is **fermented** by biological enzyme to produce PDO. The reaction is as follows. This technical preparing line has advantages of simple operation, mild circumstan, high efficiency, low cost, less by-products and

美国 DuPont 公司利用生物酶催化合成技术制备 PDO。碳水化合物通过生物酶催化发酵生成 PDO，其反应如下。该制备工艺路线具操作简便、条件温和、高效、成本低、副产物少等

so on. However, the selective range of microbial enzymes is small.

优点，但微生物酶的可选择范围较小。

$$\text{H}_2\text{C}-\text{CH}-\text{CH}_2 \xrightarrow{\text{Fermentation/发酵}} \text{H}_2\text{C}-\text{CH}_2-\text{CH}_2$$
$$\text{OH}\ \ \text{OH}\ \ \text{OH} \qquad\qquad \text{OH}\qquad\text{OH}$$

2.2 PTT Polymerization/PTT 合成

PTT is polymerized by **melt polycondensation**. The process routes mainly include direct **esterification polycondensation (TPA route)** and ester exchange polycondensation (DMT route). The difference between the two methods lies in the synthesis process of **propanediol terephthalate** and the subsequent polycondensation process is the same. The reaction is carried out in two steps: the first step is **transesterification** or esterification, and the second step is polycondensation.

In the reaction process, chemical side reactions usually occur, mainly macromolecular chain **pyrolysis** of PTT melt at high temperature, which produces two kinds of by-products: **oligomers** and volatile small molecular organics. In general, the oligomer content of PTT ranges from 1.6% to 3.2%, higher than that of PET (1.7%) and of PBT (1%). The presence of oligomers affects the filament spinning and dyeing process. Volatile small molecular organics, such as 0.2%~0.3% propionaldehyde and **allyl alcohol**, are typically removed by **distillation** method.

采用熔融缩聚法合成PTT，其工艺路线主要有直接酯化法缩聚（TPA路线）和酯交换法缩聚（DMT路线）。这两种方法的差别在于对苯二甲酸丙二醇酯的合成过程不同，而后的缩聚过程是相同的。反应分两步进行：第一步为酯交换或酯化反应，第二步为缩聚反应。

在反应过程中，都会发生化学副反应，主要是PTT熔体在高温下发生大分子链热裂解反应，生成齐聚物和挥发性小分子有机物两类副产物。一般情况下，PTT中齐聚物的含量为1.6%~3.2%，高于PET的1.7%和PBT的1.0%，这些齐聚物会影响纺丝和染色加工，挥发性小分子有机物包括0.2%~0.3%的丙醛和烯丙醇，这类副产物可用精馏方法除去。

2.2.1 Direct Esterification Method (TPA Route)/ 直接酯化法（TPA 路线）

Direct esterification method is to directly **esterify terephthalic acid** (TPA) with PDO to obtain propylene terephthalate, and then **depressurize** to carry out

直接酯化法就是对苯二甲酸（TPA）与PDO直接进行酯化反应，一步法制得对苯二甲

precondensation and polycondensation reaction to obtain PTT. Since TPA is an amorphous powder under normal condition, its melting point (425 ℃) is higher than its **sublimation** temperature (300 ℃), and boiling point of PDO (215 ℃) is lower than sublimation temperature of TPA (300 ℃), therefore direct esterification system is a **heterogeneous** system in which the solid phase TPA coexists with the liquid phase PDO. The esterification occurs only between PDO and the dissolved TPA in PDO. The TPA route is shown in Figure 7-1. The reactions are as follows.

酸丙二酯，再降压进行预缩聚和缩聚反应来制得PTT。由于TPA在常态下为无定形粉末，其熔点（425 ℃）高于其升华温度（300 ℃），而PDO沸点（215 ℃）又低于TPA的升华温度（300 ℃）。因此，直接酯化体系为固相TPA与液相PDO共存的多相体系，酯化反应只是发生在已溶解于PDO中的TPA与PDO之间，TPA路线如图7-1所示，其反应如下：

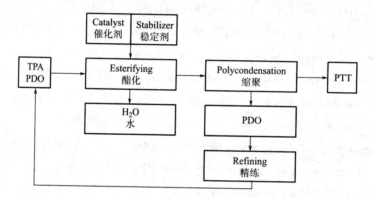

Figure 7-1　TPA route
图7-1　TPA路线

$$nHOOC-\text{C}_6H_4-COOH + 2nHO-(CH_2)_3-OH \underset{k_2}{\overset{k_1}{\rightleftharpoons}} HO-(CH_2)_3-OOC-\text{C}_6H_4-COO-(CH_2)_3-OH + 2nH_2O$$

The TPA consumed in the reaction is supplemented by the following dissolved TPA in the solution. Because of low TPA solubility in PDO, the liquid phase in the system is a TPA saturated solution before the TPA completely dissolves. Therefore, the esterification rate is not relevant to TPA concentration. The reaction equilibrium preceeds towards the direction of forming trimethylene terephthalate.

反应消耗的TPA由溶液中后溶解的TPA来补充，由于TPA在PDO中的溶解度不大，在TPA全部溶解前，体系中液相为TPA的饱和溶液，因此，酯化反应速度与TPA浓度无关，反应平衡向生成对苯二甲酸丙二酯的方向进行。

2.2.2 Ester Exchange Method(DMT Route)/ 酯交换法（DMT 路线）

PDO is exchanged with **methoxy** (–OCH$_3$) in **dimethyl terephthalate** (DMT) to form **polyethylene terephthalate** in the presence of catalyst. The substituted methoxy group combines with hydrogen in PDO to produce **methanol**. The generated polyethylene terephthalate is then polycondensed to form propylene terephthalate at 140~220 ℃ with titanium butyl or tetrabutoxy titanium as the **catalyzer**. After the removal of by-product methanol, polycondensation goes on when the temperature rises to 270 ℃ and pressure drops to 5 kPa. The DMA route is shown in Figure 7-2. The reaction is as follows. DMT route has the advantages of easy control, convenient operation, etc in laboratory. However, it is not suitable for the industrial production by DMT route due to the long process, many devices, large investment, heavy pollution and high cost.

在催化剂作用下，PDO 与对苯二甲酸二甲酯（DMT）中的甲氧基（—OCH$_3$）交换，生成对苯二甲酸双羟丙酯，被取代的甲氧基与 PDO 中的氢结合，生成甲醇。然后生成的对苯二甲酸双羟丙酯再进行缩聚反应生成对苯二甲酸丙二酯，反应温度在 140~220 ℃，催化剂为丁基化钛或者四丁氧基钛，反应后除去副产物甲醇，再将温度升高至 270 ℃，压力降到 5kPa 进行缩聚。DMA 路线如图 7-2 所示，反应如下。实验室 DMT 路线具有容易控制、操作方便等优点。但在实际生产中，流程长，设备多，投资大，污染重，成本高，因此，DMT 路线不适合于工业化生产。

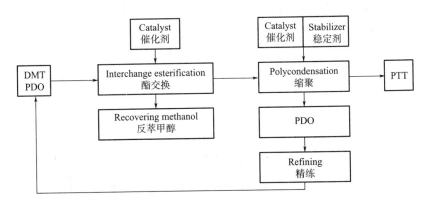

Figure 7-2 DMT route
图7-2 DMT路线

2.2.3 Polycondensation/ 缩聚反应

Polycondensation is the necessary process for both two reaction methods described above. The reaction is as follows.

缩聚反应是上述两种反应方法都必须经过的，反应如下：

$$n\text{HO}\!-\!(\text{CH}_2)_3\!-\!\text{OOC}\!-\!\!\!\bigcirc\!\!\!-\text{COO}\!-\!(\text{CH}_2)_3\!-\!\text{OH}$$
$$\longrightarrow -\!(\text{OC}\!-\!\!\!\bigcirc\!\!\!-\text{COO}\!-\!(\text{CH}_2)_3\!-\!\text{O})_n\!+n\text{HO}\!-\!(\text{CH}_2)_3\!-\!\text{OH}$$

The mechanism can be elucidated by coordination theory. The hydroxyl hydrogen in one propylene terephthalate molcule is combined with the carbonyl oxygen, and it is easy to form inner cyclic complex by itself. In the polycondensation reaction, the hydrogen atom is replaced by the metal ion (M) in the catalyst to produce the metal alcohol compound, which is the active intermediate of the polycondensation reaction. The empty orbit provided by the metal in chelate coordinates with the lone pair electrons of carboxyl oxygen to increase the positive electricity of carbonyl carbon atom. The hydroxyl oxygen on another propylene terephthalate molcule is combined with carbonyl carbon to carry out the polycondensation reaction. The reactions are as follows.

其机理可用配位理论来解释。一个对苯二甲酸丙二酯分子的羟基氢与羰基氧进行结合，很容易自身生成内环状络合物。在缩聚反应中，氢原子则被催化剂中的金属离子（M）置换，生成金属醇化合物，这就是缩聚反应中的活性中间体。螯合物中的金属提供空轨道与羧基氧的孤对电子配位，增加羰基碳原子的正电性，另一个对苯二甲酸丙二酯分子的羟基氧与羰基碳结合，从而完成缩聚反应。其反应如下：

Polycondensation process is divided into low vacuum stage and high vacuum stage. The removal of the formed small molecules such as PDO in PTT condensation process

缩聚过程分为低真空阶段和高真空阶段，在 PTT 缩聚过程中，生成的小分子如 PDO 的脱

is more difficult than that in PET production. The interface should be updated timely to remove small molecules so that the degree of polymerization can reach the requirement. A higher vacuum degree therefore is required in the poly condensation process, which needs to be strictly controlled.

2.3 Spinning/ 纺丝

The chip moisture content should generally be less than 30 μg/g in PTT melt spinning. Drying treatment therefore has to be employed to PTT chip before spinning. The drying of PTT chip is different from that of PET chip in that it only requires a multi-stage dryer rather than crystallizer.

In the extrusion section, the temperature of the PTT melt usually is 30 ℃ lower than that of PET, and the melt temperature of PTT ranges from 245 ℃ to 265 ℃. The polymer will obviously be thermally degraded when the melt temperature is above 265 ℃.

It is easy to produce roll forming and cause bobbin collapsing due to elasticity of PTT fiber. Therefore, the properly increasing winding speed can improve the forming process and 4000 m/min is preferable.

3 Structure/ 结构

Chemical structures of PET、PTT and PBT are shown in Figure 7-3. There are three **methylene** groups per chain unit in PTT molecule, which generates an "odd carbon effect" among the macromolecular chains, thus forming a spiral arrangement.

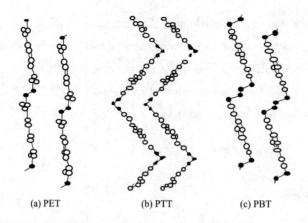

(a) PET　　　(b) PTT　　　(c) PBT

Figure 7-3　Chemical structures of PET, PTT and PBT
图7-3　PET、PTT和PBT的化学结构式

4　Property/ 性能

Elasticity: Elastic recovery curves of PTT and other fibers are shown in Figure 7-4. PTT fiber has excellent elastic recovery. Elastic recovery of PTT fiber is almost two times that of PET fiber. It can recover to its original length when stretched by 20%. It still recovers to almost 100% of its original length even stretched by 20% after 10 times.

弹性：PTT 和其他纤维的弹性回复曲线如图 7-4 所示。PTT 纤维具有优异的弹性回复性能。其弹性回复性几乎是 PET 纤维的两倍。即使拉伸 20% 仍可回复至原长，经过 10 次 20% 的拉伸仍然能几乎 100% 的回复。

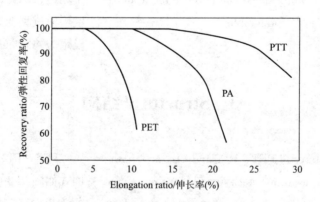

Figure 7-4　Elastic recovery ratio of PTT and other fibers
图7-4　PTT和其他纤维弹性回复率

Dyeability: PTT fiber has a better dyeability than PET fiber considering that glass transition temperature of PTT

染色性：PTT 纤维的玻璃化转变温度比 PET 纤维大约低

fiber is about 20 ℃ lower than that of PET fiber. That is to say, PTT fiber can be darkly dyed with low temperature disperse dyes at room temperature and atmospheric pressure, and it has good color fastness.

20 ℃，所以PTT纤维的染色性能优于PET纤维。即使在常温常压下用低温型分散染料也能染成深色，而且具有较好的色牢度。

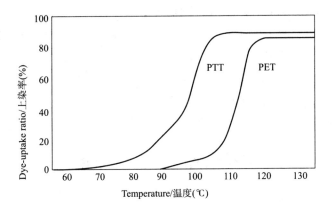

Figure 7–5 Dye-uptake ratio of PTT and PET fibers
图7-5 PTT和PET纤维上染率

Handle and drapability: Low modulus of PTT fiber provides a soft and drapable fabric.

手感和悬垂性：PTT纤维模量较低，其织物具有柔软的手感和良好的悬垂性。

5 Application/ 应用

PTT fiber is suitable for making carpet due to its outsanding bulkiness and resilience as well as excellent stain resistance (especially after proper chemical finishing). PTT filament and staple fiber can be processed into various knitted and woven fabrics with high elasticity and soft handle, which has PAN-like bulkiness, PET-like antifouling and PA-like softness.

PTT纤维蓬松性和回弹性好，防污性又十分优良（尤其经适当的化学整理后），因此，PTT纤维比较适合于制作地毯。PTT长丝和短纤可用于生产各种高弹及手感柔软的针织物和机织物，具有腈纶般的蓬松性，PET的防污性及锦纶的柔软性。

Exercises/ 练习

1. What is the definition of polyester fiber?
2. Describe chemical structure of PET, PBT and PTT.
3. Describe main property of PET.
4. Describe the maufacturing method of PTT.
5. Describe structure and property of PTT.
6. Describe the application of PTT.

References/ 参考文献

［1］肖雪春，李文刚，黄象安. PET/PTT 共混体系的非等温结晶动力学研究 [J]. 合成纤维，2005，11: 10-14.

［2］吕初旭，柴焕敏. PTT 的结构及性能［J］. 聚酯工业，2004，17（4）: 6-10.

［3］吴惠英. PTT 及其与 PET 和 PBT 共混纤维的性能分析［J］. 现代纺织技术，2008，5: 8-10.

［4］任永花，俞建勇，张一心. PTT 纤维及其制品的应用特性分析［J］. 棉纺织技术，2005，33（11）: 1-4.

［5］武振侠，肖刚，郑帼. 国内外 PTT 现状及发展建议［J］. 聚酯工业，2014，27(5): 1-3.

［6］CHEN M Y, LAI K, SUN R J, et al. Effects of drawing technology on the properties of PTT/PET bi-component filaments［J］. 西安工程大学学报，2009，23（2）: 535-540.

［7］乔宇. PTT 及 PTT—PEG 嵌段共聚物的合成和相转变的研究［D］. 北京：北京服装学院，2009.

Part Four

High Performance Fiber/ 高性能纤维

Chapter Eight　Carbon Fiber/ 碳纤维

1　Introduction/ 引言

Carbon fiber (CF) is a high performance fiber with more than 90% carbon content. It is made of microcrystalline **graphite** prepared by **stack**ing **flake** graphite **microcrystal**s along the fiber axial direction, and then **carboniz**ing and **graphitiz**ing. As a new type of carbon material, CF has been developing continuously for more than 150 years since its inception. CF was first applied as luminescence of incandescent lamp in the United States. Now, CF has entered many aspects of people's lives. It can be seen everywhere, ranging from mobile phone shell to space shuttle. The **feedstock** of CF has developed from Rayon to polyacrylonitrile (PAN), **isotropic** and **mesophase picth, hydrocarbon** gas and **ablate**d graphite. Rayon-based CF has no longer been produced, and PAN-based CF dominates.

Dreselhaus, Donnet and Pebles et al. published some books on CF. The physical property, development from 1980 to 1992 and technology in the mid-1980s of CF were reviewed, respectively.

碳纤维是一种含碳量在90%以上的高性能纤维，由片状石墨微晶沿纤维轴向方向堆砌，经碳化及石墨化处理而得到的微晶石墨材料。作为新型碳材料，碳纤维从问世到现在已经有150多年了，一直在不断发展。碳纤维最早作为白炽灯的发光体诞生于美国，目前已进入到人们生活的许多方面，随处可见，小到手机壳，大到航天飞机。碳纤维原料从黏胶纤维发展到聚丙烯腈（PAN）、各向同性和中间相沥青、碳氢气体和烧蚀石墨。黏胶基碳纤维已不再生产，PAN基碳纤维占主导地位。

Dreselhaus、Donnet 和 Peebles 等出版了关于碳纤维的书，分别总结了碳纤维的物理特性、发展（1980~1992）以及技术（80年代中期）。

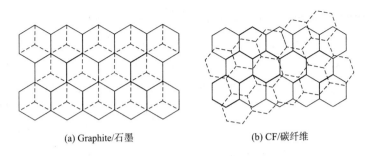

(a) Graphite/石墨　　(b) CF/碳纤维

Figure 8–1　Stacking of aromatic sheets
图8-1　芳香族片的堆积

The stacking of **aromatic** sheets in graphite and CF are shown in Figure 8–1. Although CF is often referred to as graphite fiber, the aromatic sheets in CF randomly arrange described as "**turbostratic**", which do not regularly stack as that in graphite. Large aromatic sheets determine many physical properties of CF.

CF can be classified by precursor, mechanical property, treatment temperature, application field, fiber length and function. According to **precursor**, CF is divided into PAN-based CF, pitch-based CF, viscose-based CF and vapor-grown CF. Depending on mechanical property, CF is classified into general grade (GP) CF and high performance (HP) CF. HPCF includes medium tenacity (MT), high tenacity (HT), ultra-high tenacity (UHT), medium modulus (IM), high modulus (HM) and ultra-high modulus (UHM). Based on treatment temperature, CF is separated into anti-inflammatory CF, carbon CF and graphite CF. The treatment temperature of anti-inflammatory CF is 200~250 ℃, which can be used as electrical insulator. The treatment temperature of carbon CF is 500~150 ℃, which can be used as electrical conductive material. The treatment temperature of graphite CF is above 2000 ℃. Its thermal resistance and electrical conductivity are improved and it also has **lubricity**. In the light of application field, CF contains commodity grade CF and aerospace grade CF. On the basis of length, CF is partitioned into cotton-like staple

芳香片层在石墨和碳纤维中的堆积如图 8-1 所示。虽然碳纤维通常被称为石墨纤维，但在碳纤维中，芳香片层不像石墨中规则堆积，而是随机排列，称为"乱层"。大的芳香片层决定碳纤维的许多物理性质。

碳纤维可以根据前驱体、力学性能、处理温度、应用领域、纤维长度和功能来分类。根据前驱体分类，碳纤维分为 PAN 基碳纤维、沥青基碳纤维、黏胶基碳纤维和气相生长碳纤维。根据力学性能分类，碳纤维分为普通级碳纤维（GPCF）和高性能碳纤维（HPCF），其中高性能碳纤维包括中强型（MT）、高强型（HT）、超高强型（UHT）、中模型（IM）、高模型（HM）、超高模型（UHM）。根据处理温度分类，碳纤维分为耐炎质碳纤维、碳素质碳纤维和石墨质碳纤维。耐炎质碳纤维处理温度为 200~250 ℃，可做电气绝缘体；碳素质碳纤维处理温度为 500~1500 ℃，可用作导电材料；石墨质碳纤维处理温度在

fiber and **filamentous** continuous fiber. In terms of function CF is partitioned into forced structural CF, fire resistant CF, abrasion resistant CF, corrosion resistant CF, active CF, electrical conductive CF and CF for **lubrication**.

2000 ℃以上，除耐热性与导电性提高外，还具润滑性。根据应用领域分类，碳纤维分为商品级和宇航级。根据长度分类，碳纤维分为棉状短纤维和长丝状连续纤维。根据功能分类，碳纤维可分为受力结构碳纤维、耐火碳纤维、耐磨碳纤维、耐腐蚀碳纤维、活性碳纤维、导电碳纤维和润滑用碳纤维。

2　Preparation/制备

The **carbonization** yield of PAN-based CF is higher than that of viscose-based CF, up to 45%. The production process, solvent recovery and three wastes treatment for PAN-based CF are simpler than those for viscose-based CF. In addition raw material is abundant. The cost is low and the mechanical properties are excellent. Especially the tensile tenacity and modulus rank first among the three kinds of CFs. Therefore, it is now the most widely used CF with the largest output. The preparation process of PAN-based CF as shown in Figure 8-2 includes five steps: pre-oxidation, carbonization, **graphitization**, **sizi**ng and surface treatment.

Under certain polymerization condition, the double bonds of **acrylonitrile** (AN) are opened under the action of free radicals of initiator and linked to each other as linear polyacrylonitrile (PAN) macromolecular chain. PAN fiber is made from PAN spinning solution by wet spinning. PAN fiber is pre-**oxidize**d in pre-oxidizing furnace to obtain pre-oxidized fiber, which is then carbonized into high tenacity CF in carbonization furnace. High tenacity CF is graphitized into high modulus CF. Finally, the sizing and

PAN基碳纤维的碳化收率比黏胶基碳纤维高，可达45%，其生产流程、溶剂回收和三废处理都比黏胶基碳纤维简单。另外，原料来源丰富，成本低，力学性能优异，尤其是拉伸强度和拉伸模量居三种碳纤维之首。所以，PAN基碳纤维是目前应用领域最广、产量最大的一种碳纤维。PAN基碳纤维制备过程如图8-2所示，分为五步：预氧化、碳化、石墨化、上浆及表面处理。

在一定的聚合条件下，丙烯腈（AN）在引发剂的自由基作用下，双键被打开，并彼此连接为线型聚丙烯腈（PAN）大分子链。PAN纺丝液经过湿法纺丝得到PAN纤维，经预氧化炉预氧化后得到预氧化纤维，经碳化炉碳化后得到高强碳纤维，再经石墨化后得到高模碳纤维，最后

surface treatment of CF is carried out. Sizing is conducted to protect CF, and the sizing agents are served as a coupling agent between CF and resin. CF is often treated with strong oxidants (such as concentrated sulfuric acid, concentrated nitric acid, hypochlorite and dichromate) and oxygen-containing groups are introduced on the surface to improve the adhesion of CF.

对碳纤维进行上浆和表面处理。上浆是为了保护碳纤维，另外浆料作为碳纤维与树脂的偶联剂。常利用强氧化剂（如浓硫酸、浓硝酸、次氯酸、重铬酸）处理碳纤维，表面引入含氧基团，以提高碳纤维表面的黏结性能。

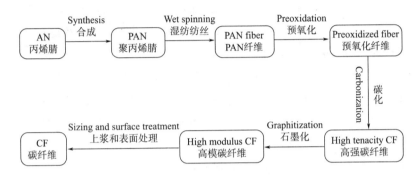

Figure 8-2 Preparation process of PAN-based CF
图8-2 PAN基碳纤维制备过程

The glass transition temperature (T_g) of **homopolymer** PAN is 104 ℃, and there is no softening temperature. It decomposes at 317 ℃. The T_g of **copolymer** PAN is about 85—100 ℃ and changes with copolymer component and content. The higher the copolymer content is, the lower the T_g is. The linear macromolecular chain of **thermoplastic** PAN is transformed into non-plastic and thermal resistant **trapezoidal** structure, which results in no melting and no burning at high carbonization temperature, maintenance of fiber morphology, and stable thermodynamic state. The pre-oxidation temperature is controlled between the glass transition temperature and the **decomposition** temperature, i.e. 200—300 ℃. The trapezoidal structure by pre-oxidation greatly improves the carbonization efficiency and reduces the production cost. At the same time, the pre-oxidized fiber is also an important intermediate product and can be further processed into a variety of products, which

均聚 PAN 的玻璃化转变温度（T_g）为 104 ℃，没有软化温度，在 317 ℃ 分解。共聚 PAN 的 T_g 为 85~100 ℃，T_g 随共聚物组分和含量变化而变化。共聚含量越高，T_g 越低。热塑性 PAN 线型大分子链转化为非塑性耐热梯形结构，使其在碳化高温下不熔不燃、保持纤维形态，热动态稳定。预氧化的温度控制在玻璃化温度和分解温度之间，即 200~300 ℃。预氧化的梯形结构使碳化效率显著提高，大大降低了生产成本。同时，预氧化纤维也是一种重要的中间产品，可进一步加工成多种产品，直接进入市场，并已在许多领域得到实

directly enter the market, and has been practically applied in many fields.

After pre-oxidation treatment PAN precursor is transformed into thermal resistant trapezoidal structure, which is transformed into CF with **disorder**ed graphite structure after low temperature carbonization (300~1000 ℃) and high temperature carbonization (1000~1800 ℃). In the process of structural transformation, smaller trapezoidal structural units are further crosslinked and condensed. Many small molecule by-products are released. At the same time, non-carbon elements O, N, H are gradually removed and C is gradually enriched. Finally high tenacity CF with more than 90% carbon content is formed. High modulus graphite fiber is formed by heat treatment of high tenacity CF at 2000~3000 ℃. The amorphous and disordered graphite structure is transformed into three-dimensional graphite structure.

For CF the preoxidation time is nearly 100 minutes, the carbonization time is several minutes, and the graphitization time is relatively short, usually only a few seconds to tens of seconds.

际应用。

PAN原丝经预氧化处理后转化为耐热梯形结构，再经过低温碳化（300~1000 ℃）和高温碳化（1000~1800 ℃）转化为具有乱层石墨结构的碳纤维。在这一结构转化过程中，较小的梯形结构单元进一步进行交联、缩聚，释放出许多小分子副产物。同时，非碳元素O、N、H逐步被去除，C逐渐富集，最终形成含碳量90%以上的高强度碳纤维。在2000~3000 ℃热处理高强度碳纤维，由无定型、乱层石墨结构向三维石墨结构转化，形成高模量石墨纤维。

对于碳纤维来说，预氧化时间为近百分钟，碳化时间为几分钟，石墨化时间较短，一般只有几秒到数十秒。

3 Structure/ 结构

3.1 Sheath-Core Structure Model/ 皮芯结构模型

Sheath-core structure model, proposed by Bennett and Johnson, believed that the apparent sheath-core structure can be observed in CF. The cortex is denser and the core with many micropores is looser. Graphite microcrystallites are well oriented along the fiber axis in sheath, and present creasing **pleat** in core. Meanwhile complex micropores are observed between graphite layers. In spinning process, the filament coagulation shows the process of double diffusion. It begins on the filament surface, and the membrane is then

Bennett和Johnson提出的皮芯结构模型认为碳纤维具有明显的皮芯结构，皮层较致密，芯部较为疏松，有很多微孔。石墨微晶在皮层沿着纤维轴向有序排列，在芯部呈现褶皱形态，同时，石墨片层之间存在错综复杂的微孔。纺丝过程中，丝条凝固是一个双扩散过程，首先在丝条

formed on the surface, which is unfavorable for diffusion into core. There is difference in densification degree between surface and interior of precursor. And the sheath-core structure is formed as a result.

表层进行，表面形成薄膜，这不利于扩散向芯部进行，导致原丝表层和内部致密化程度存在差异，形成皮芯结构。

3.2 Microfibril Model/ 微原纤模型

Microfibril mode, put forward by Difendorf and Tokarsky is that the basic unit of CF is the microfibril, which is composed of 10~30 basic carbon networks. The microfibril is well aligned with the fiber axis and stacked into belt structure with small gaps between layers, large layer stack and adequate proportion of crystal orientation.

Diefendorf 和 Tokarsky 提出的微原纤模型认为碳纤维的基本单元是微原纤，该微原纤由 10~30 个基本碳网面构成，沿着纤维轴向取向排列，堆叠成条带结构，层面间距小，层面堆积大，晶向比例大。

4 Property/ 性能

4.1 Mechanical Property/ 力学性能

CF has high tenacity, high modulus and low density (1.6~2.5 g/cm^3). Therefore, it has high specific tenacity and modulus, which are 7 and 5 times most metals, respectively. Young's modulus of CF is more than three times that of glass fiber and about two times that of Kevlar fiber. The mechanical properties of CFs with different precursors and with different tenacity and modulus are shown in Table 8-1 and Table 8-2, respectively.

碳纤维的强度高，模量大，密度小（1.6~2.5 g/cm^3），因此具有较高的比强度和比模量。其比强度和比模量分别是绝大多数金属的 7 倍和 5 倍。碳纤维杨氏模量是玻璃纤维的 3 倍多，是凯夫拉纤维的 2 倍左右。不同前驱体、不同强度和模量的碳纤维力学性能分别见表 8-1 和表 8-2。

Table 8-1 Mechanical properties of CFs with different precursors
表 8-1 不同前驱体的碳纤维的力学性能

Properties 性能	PAN-based CF PAN基碳纤维	Pitch-based CF 沥青基碳纤维	Viscose-based CF 黏胶基碳纤维
Tensile tenacity 拉伸强度（GPa）	2.5~3.1	1.6	2.1~2.8

续表

Properties 性能	PAN-based CF PAN基碳纤维	Pitch-based CF 沥青基碳纤维	Viscose-based CF 黏胶基碳纤维
Tensile modulus 拉伸模量（GPa）	207~345	379	414~552
Density 密度（g/cm³）	1.8	1.7	2.0
Breaking elongation ratio 断裂伸长率（%）	0.6~1.2	1	—

Table 8-2　Mechanical properties of CFs with different tenacity and modulus
表 8-2　不同强度和模量的碳纤维的力学性能

Properties 性能	UHM 超高模型	HM 高模型	UHT 超高强型	HT 高强型	IM 中模型
Tensile tenacity 拉伸强度（GPa）	>400	300~400	200~350	200~250	180~200
Tensile modulus 拉伸模量（GPa）	>1.7	>1.7	>2.76	2.0~2.75	2.70~3.0
Carbon content 碳含量（%）	99.8	99.0	96.5	96.5	99.0

The modulus of CF increases with the increase of treatment temperature. When the temperature increases, the crystallization zone increases and the regular area of carbon ring arrangement enlarges, which results in the increase of crystallinity and orientation. Mechanical properties of CFs with different treatment temperature areshown in Table8-3. CF treated at 2500 ℃ is called high modulus CF (type I CF), also known as graphite fiber. Its elastic modulus is 400—600 GPa, and the elongation at break is about 0.5%. CF treated at 1300—1700 ℃ is called high tenacity CF (type II CF). Its tenacity is the highest. The cost of CF increases with the increase of treatment temperature.

碳纤维的模量随着处理温度的升高而提高，温度升高，结晶区增长，碳环排列规整区域扩大，使结晶度和取向度提高。不同处理温度处理的碳纤维的力学性能见表 8-3。2500 ℃处理的碳纤维称为高模量碳纤维（Ⅰ型碳纤维），又称石墨纤维，弹性模量为 400~600 GPa，断裂伸长率约为 0.5%。1300~1700 ℃处理的碳纤维称为高强度碳纤维（Ⅱ型碳纤维），此时，碳纤维强度最

Table 8-3　Mechanical properties of CFs with different treatment temperature
表 8-3　不同处理温度处理的碳纤维的力学性能

Properties 性能	Type A (Common, Moderate tenacity) 品种A（常规，中等强度）	Type II Ⅱ型	Type I Ⅰ型
处理温度 Treatment temperature(℃)	1000	1500	2500

续表

Properties 性能	Type A (Common, Moderate tenacity) 品种A（常规，中等强度）	Type II II型	Type I I型
Density 密度（g/cm³）	1.73	1.75	1.90~2.0
Tensile tenacity 拉伸强度（GPa）	2.1	2.5	1.4~2.1
Tensile modulus 拉伸模量（×10²GPa）	2	2.2~2.5	3.9~4.6
Breaking elongation ratio 断裂伸长率（%）	1.2~1.5	1.0	0.5

CF suffers less due to abrasion resistance when it is ground with metal. CF has stable structure, no **creep**, high **fatigue** tenacity and good fatigue resistance. After millions of cyclic stress fatigue tests, the tenacity retention ratio of CF reinforced composite is still 60%. However, those of steel, **aluminum** and glass fiber reinforced composite are 40%, 30% and 20%~25%, respectively.

CF is very **brittle**. It has low resistance to pressure, **shear** and **impact**, etc. The stressed deformation of CF is small due to high modulus. There is no obvious deformation sign before fiber breaking, and CF exhibits brittle fracture. It is therefore difficult to predict the **failure** of CF.

4.2 Thermal Property/ 热学性能

CF is prone to be oxidized into carbon dioxide due to its poor **oxidative** resistance when the temperature exceeds 400 ℃. 30% **phosphoric acid** (H_3PO_4) is generally applied to treat CF so that the oxidative resistance can be improved. It can withstand ultra-high temperature in non-oxidizing environment. The properties are very stable below 400 ℃ and do not change much even at 1000 ℃. It still

高。碳纤维成本随着处理温度的升高而增加。

碳纤维耐磨，与金属对磨时，磨损较少。碳纤维结构稳定，无蠕变，疲劳强度高，耐疲劳性好。经数百万次的应力疲劳循环试验后，碳纤维增强复合材料强度保留率仍有60%，而钢材、铝材和玻璃纤维增强复合材料的强度保留率分别为40%、30%、20%~25%。

碳纤维较脆，耐压、耐剪切、抗冲击等性能较差。碳纤维模量高，受力后变形较小，碳纤维断裂前无明显的形变征兆，发生脆断，因此，较难预测碳纤维的破坏。

碳纤维耐氧化性能较差，当温度超过400 ℃时容易被氧化成二氧化碳。可利用30%的磷酸处理碳纤维，提高其耐氧化性能。非氧化环境下碳纤维耐超高温，在400 ℃以下性能非常稳定，甚至在1000 ℃时无太大变

has tenacity at 2000 ℃ and no **brittle fracture** occurs in liquid nitrogen.

CF has good thermal conductivity along fiber axis, which is close to that of steel. However, the thermal conductivity decreases with the increase of temperature. The linear expansion coefficient of CF has a negative temperature effect along fiber axis. CF tends to shrink with the increase of temperature.

化，2000 ℃时仍有强度，液氮中也不脆断。

碳纤维沿着纤维轴向的导热性较好，接近于钢铁，但其导热性随温度升高而降低。碳纤维的线膨胀系数沿纤维轴向具有负温度效应，随温度升高，碳纤维有收缩的趋势。

4.3 Other Properties/ 其他性能

Except strong oxidizing acid, CF does not dissolve and swell in organic solvents, acids and alkalis, and has good **corrosion resistance**. The volume and mechanical properties slightly change after soaking in sodium hydroxide solution. It has good electrical conductivity and **electromagnetic** shielding property along fiber axis, and its electrical conductivity is between nonmetal and metal. It has self-lubricating property due to small friction coefficient. X-ray transmission of CF is good.

强氧化酸以外，碳纤维在有机溶剂、酸、碱中不溶不胀，具有较好的耐腐蚀性。在氢氧化钠溶液中浸泡后体积和力学性能变化较小。碳纤维沿纤维轴向具有良好的导电和电磁屏蔽性能，其导电性介于非金属和金属之间。碳纤维摩擦系数小，具有自润滑性能。X射线透过性好。

5 Application/ 应用

CF has many excellent properties such as low density, high tenacity, high modulus, abrasion resistance, high temperature resistance, corrosion resistance, thermal conductivity, electrical conductivity and low thermal expansion. Therefore it has been widely used in aerospace, construction, military and sports industries. CF composite can be widely used in aerospace, automobile industry, sports equipment and so on due to light weight, high tenacity and high modulus. CF reinforced resin composite is an important shell material for aircraft, warship, spaceship, rocket, missiles. CF reinforced amorphous carbon composite is an important brake disc material for missile, rocket,

碳纤维具有低密度、高强度、高模量、耐磨、耐高温、耐腐蚀、导热、导电以及低热膨胀等优异性能，已被广泛应用于航空航天、建筑、军事和体育等行业。尤其轻质、高强度、高模量，碳纤维复合材料可广泛应用于航空航天、汽车工业、运动器材等。碳纤维增强树脂复合材料是飞机、舰艇、宇宙飞船、火箭、导弹的重要壳体材料。碳纤维增强无定形碳复合材是导弹、

fire hose and aircraft. CF is applied to replace asbestos to make high-grade rubbing materials owing to abrasion resistance, which have been used as brake pads for aircraft and automobile. It can be used as solar collector material and heat conduction shell material with uniform heat conduction because of thermal conductivity. The expansion coefficient of the composite made from it is naturally stable and can be used as standard measuring instrument. It can be served as structural material of submarine on account of outstanding damping and excellent penetration sonar, such as sonar dome in submarine. Conductive material and electromagnetic shielding material are achieved due to the electrical conductivity of CF, which has been used in great quantities in communication, construction, civil engineering and so on. High X-ray transmittance has been found an increasingly wide utilization in medical equipment.

火箭、喷火喉管及飞机等重要刹车盘材料。因耐磨性好，可用碳纤维来取代石棉制备高级的摩擦材料，已作为飞机和汽车的刹车片材料。导热性好，可作为太阳能集热器材料、传热均匀的导热壳体材料。热膨胀系数小，由它制成的复合材料膨胀系数自然比较稳定，可作为标准测量器具。突出的阻尼与优良的透声纳，可作为潜艇的结构材料，如潜艇的声纳导流罩等。利用碳纤维导电特性而制成导电材料和电磁屏蔽材料，在通信、建筑、市政工程等有广泛的应用。高X射线透射已经在医疗器材中得到广泛应用。

Exercises/ 练习

1. What are the outstanding properties of carbon fibers? Why does the fiber exhibit the performances?

2. Translate the following Chinese into English.

（1）碳纤维比金属铝轻，但强度高于钢铁，具有耐腐蚀、高模量的特性，在国防军工和民用方面都是重要材料。

（2）皮芯结构模型认为碳纤维具有明显的皮芯结构，表皮较致密，芯部较为疏松，有很多微孔。

（3）碳纤维暴露在空气中，随着温度的升高（200~290 ℃开始）会发生氧化反应，当温度超过400 ℃，会出现明显的氧化反应，并以CO和CO_2的形式从表面散失。

（4）碳纤维较脆，耐压、耐剪切能力较差，抗冲击能力较差。耐氧化性能较差，高温下容易被氧化成二氧化碳。

（5）碳纤维与无定形碳的复合材料耐高温和耐烧蚀，是导弹、火箭、喷火喉管及飞机等刹车盘的重要制造材料。

References/ 参考文献

[1] NEWCOMb B A. Processing, structure, and properties of carbon fibers[J]. Composites Part A: Applied Science and Manufacturing, 2016, 91: 262-282.
[2] HEARLE J W S. High-performance fibers[M]. Cambridge: Woodhead Publishing Limited, 2001.
[3] 晏雄. 产业用纺织品[M]. 上海：东华大学出版社，2010.
[4] 张以河. 复合材料学[M]. 北京：化学工业出版社，2011.

Chapter Nine   Glass Fiber/ 玻璃纤维

1 Introduction/ 前言

Glass fiber (GF) is a kind of inorganic non-metallic material with excellent properties. It is made from pyrophyllite, **quartz** sand, limestone, dolomite, borocalcite and brucite by high temperature melting, drawing, etc. The diameter of monofilament ranges from several microns to more than twenty microns, equivalent to 1/20~1/5 of a hair strand. The multifilament is made of hundreds or even thousands of monofilaments.

玻璃纤维（玻璃纤维）是一种性能优异的无机非金属材料，它是以叶腊石、石英砂、石灰石、白云石、硼钙石和硼镁石为原料经高温熔制、拉丝等工艺制成的，单丝直径为几微米到二十几微米，相当于一根头发丝的1/20~1/5，数百根甚至上千根单丝组成复丝。

1.1 GF Development/ 玻璃纤维发展概况

As early as the 1960s, GF was used in aircraft, but due to the high price and poor technological performance at that time, it could not get further development and attention. Later, with the improvement of technology and the expansion of application field, GF is more and more applied in military field, especially accounting for 67.7% of the reinforced fiber used in aerospace and aviation. Subsequently, its application scope is expanding day by day, and GF and its composites have been popularly used in many fields such as sports equipment, building component, light industrial product, chemical pipeline, automotive industry, medical equipment, boat and ship, etc. Since the 1980s, its annual average growth ratio has reached about 10%.

早在20世纪60年代，玻璃纤维在飞机上就获得了应用，但由于当时的价格昂贵，工艺性能欠佳等原因，未能获得进一步的发展和重视。后来，随着技术的改进和应用领域的扩大，玻璃纤维越来越多地用于军事方面，特别是其67.7%的增强纤维用于航天、航空工业。随后，其应用范围日益扩大，如体育器具、建筑构件、轻工制品、化工管道、汽车工业、医疗器械、舟艇船舰等许多领域都已普遍采用玻璃纤维及其复合材料。自20世纪80年代以来，其年均增长率高达10%左右。

1.1.1 International Development/ 国际发展

More than 2000 years ago, ancient Egyptians manually drew GF as decorative material. In the 1930s, Americans invented the technology of **platinum crucible** drawing to prepare GF. In 1938, Owens Corning Company was established in USA, which produced GF for insulation material on an industrial scale. After World War II, the industrial production of GF became widespread and companies such as St. Gobain (France), Pilkington (UK) and Nittobo (Japan) were founded and started to produce GF. From 1959 to 1960, Owens Corning Company and PPG Company successively built GF furnace.

2000多年前，古埃及手工拉制玻璃纤维做装饰材料。20世纪30年代，美国发明了制备玻璃纤维的铂坩埚拉制工艺技术。1938年美国成立欧文斯康宁公司，用于绝缘材料的玻璃纤维生产形成工业化规模。第二次世界大战后，玻璃纤维工业化生产普及，成立了圣哥本（法国）、皮尔金顿（英国）和日东纺（日本）等公司并开始生产玻璃纤维。1959~1960年，欧文斯康宁公司和PPG公司相继建成了玻璃纤维池窑。

1.1.2 Domestic Development/ 国内发展

In late 1950s, the GF industry started in China and GF was produced with **pottery** crucible method and **nickel chromium** alloy orifice plate method. In the middle of 1960s, the platinum-substitute crucible process was successfully invented. In the 1990s, the furnace method became popular in China. GF production was started in 1958, with a capacity of 500 tons and an output of 106 tons. An industrial system was formed in 1978, with an output of 41,000 tons, ranking seventh in the world. In 1998, the GF output increased to 164,000 tons, with an annual growth ratio of 7.2%, and the drawing proportion of advanced furnace was 12%. After entering the new century, with the rapid development of GF furnace drawing process, the GF output reached 1.6 million tons in 2007, and China became the largest country of GF production capacity in the world. In 2008, the GF output in China reached 2.35 million tons. From 1998 to 2008, GF output grew at an average annual

20世纪50年代后期，我国玻璃纤维工业诞生，采用陶土坩埚法和镍铬合金孔板法生产，20世纪60年代中期，成功研制代铂坩埚法生产工艺，20世纪90年代初，池窑法开始在中国普及。1958年开始生产，当年产能500吨，产量106吨。1978年形成工业体系，产量4.1万吨，居世界第7位。1998年产量增加到16.4万吨，年增长率为7.2%，先进池窑的拉丝比例为12%。进入新世纪以后，随着玻璃纤维池窑拉丝工艺的迅速发展，我国玻璃纤维产量2007年达到160万吨，成为世界玻璃纤维产能第一大国。2008年，我

ratio of 30.5%.

国玻璃纤维产量达到 235 万吨，从 1998 年到 2008 年，玻璃纤维产量年平均增长率达到 30.5%。

1.2 GF Classification/ 玻璃纤维分类

GF can be classified by alkaline **oxide** (R_2O) content, diameter, appearance and function. According to alkaline content GF is divided into **alkali-free** GF, low-alkali GF, medium-alkali GF, high-alkali GF and special GF. Alkali-free GF, also known as E GF, contains less than 0.5% R_2O, and has good chemical stability and electrical insulation, and high tenacity. The R_2O content of low-alkali GF is less than 2%, and the chemical stability, electrical insulation and tenacity are slightly worse than those of alkali-free GF. Alkali-free GF and low-alkali GF are mainly composed of aluminoborosilicate. They are mainly used as electrical insulating materials, reinforcing materials of **glass fiber reinforced plastics** (GFRP) and tire cords. They are suitable for occasions with special requirements for acid and alkali. Medium-alkali GF, also known as C GF, has about 12% R_2O content and acceptable chemical stability. It cannot be used as electrical insulation materials because of its high alkali content. However, it has low cost and wide application. It is generally used as substrate of **latex cloth** and **window screening**, and acid **filter cloth**, etc. It can also be used as GFRP reinforcing materials with low requirements for electrode performance and tenacity. High-alkali GF, also known as A GF, contains no less than 15% R_2O. They are made from broken glass and can be used as water-proof and moisture-proof material such as **battery isolator**, pipe wrapping cloth and felt sheet. Special GF consists of pure aluminum, **magnesium** and **silicon**, also called S GF. Among them, **Mg-Al-Si** system is high tenacity and high elasticity GF, and Si-Al-**Ca**-Mg system is chemical corrosion resistant GF. It also includes lead-

玻璃纤维可按碱性氧化物含量（R_2O）、直径、外观和功能进行分类。按碱含量，玻璃纤维分为无碱玻璃纤维、低碱玻璃纤维、中碱玻璃纤维、高碱玻璃纤维和特种玻璃纤维。无碱玻璃纤维又称 E 玻璃纤维，R_2O 含量小于 0.5%，化学稳定性和电绝缘性好，强度高。低碱玻璃纤维的 R_2O 含量小于 2%，化学稳定性、电绝缘性和强度略差于无碱玻璃纤维。无碱玻璃纤维和低碱玻璃纤维的主要成分是铝硼硅酸盐，主要用作电绝缘材料、玻璃钢的增强材料以及轮胎帘子线，适合于对酸碱有特殊要求的场合。中碱玻璃纤维又称 C 玻璃纤维，R_2O 含量为 12% 左右，化学稳定性尚可，因含碱量较高，不能作为电绝缘材料。但成本较低，用途较广，一般作为乳胶布和窗纱的基材、酸性过滤布等，也可用作对电极性能和强度要求不高的玻璃钢增强材料。高碱玻璃纤维又称 A 玻璃纤维，R_2O 含量不小于 15%，由碎玻璃为原料制成，可用作蓄电池的隔离片、管道包扎布和毡片等防水防潮材料。特种玻璃纤维是由纯铝、镁、硅三元组成，又称 S 玻璃纤维，其中

containing GF, high GF and quartz fiber, etc.

Depending on fiber diameter, GF can be separated into coarse GF, primary GF, intermediate GF, advanced GF (textile GF) and ultrafine GF. Their diameters are shown in Table 9-1. In general, GF with diameter of 5~10 μm can be processed into textiles, and GF with diameter of 10~14 μm is used as **twistless roving**, non-woven fabric and fiber **felt.**

镁铝硅系是高强高弹玻璃纤维，硅铝钙镁系是耐化学腐蚀玻璃纤维。还包括含铅玻璃纤维、高硅氧玻璃纤维和石英纤维等。

按纤维直径，玻璃纤维分为粗玻璃纤维、初级玻璃纤维、中级玻璃纤维、高级玻璃纤维（纺织玻璃纤维）及超细玻璃纤维，其直径见表9-1。通常，直径为5~10 μm 的玻璃纤维可加工成纺织品，10~14 μm 的玻璃纤维用作无捻粗纱、非织造布和纤维毡。

Table 9-1　Diameter of GF with different fineness
表 9-1　不同细度玻璃纤维的直径

Fiber 纤维	Coarse GF 粗玻璃纤维	Primary GF 初级纤维	Intermediate GF 中级玻璃纤维	Advanced GF 高级玻璃纤维	Super-fine GF 超细玻璃纤维
Diameter 直径（μm）	30	20~30	10~20	3~10	<4 mm

On the basis of appearance GF can be split into continuous GF, fixed length GF and glass cotton. Continuous GF is infinitely long, also known as textile fiber. After textile processing, it can be made into glass yarn, rope, cloth, belt, twistless roving and its products. The length of fixed length GF is 300—500 mm, which can be made into felt sheets. Glass cotton is also a kind of fixed length GF, less than 150 mm in length. It is fluffier like cotton **batt**, also known as short cotton, and mainly used as insulation, sound absorption material. Glass cotton with the diameter less than 3 μm is called superfine cotton.

Judging by function GF includes high tenacity GF, high modulus GF, high temperature resistant GF, alkali resistant GF, acid resistant GF, common GF (alkali-free GF and medium-alkali GF), optical GF, low dielectric constant

按外观，玻璃纤维可分为连续玻璃纤维、定长玻璃纤维和玻璃棉。连续玻璃纤维的长度无限长，又称纺织纤维，经纺织加工后可以制成玻璃纱、绳、布、带、无捻粗纱及其制品。定长玻璃纤维的长度为300~500 mm，可制成毡片；玻璃棉也是一种定长玻璃纤维，长度小于150 mm，较为蓬松，似棉絮，又称为短棉，主要用作保温、吸声材料。直径小于3 μm 的玻璃棉称为超细棉。

按功能，玻璃纤维包括高强玻璃纤维、高模量玻璃纤维、耐高温玻璃纤维、耐碱玻璃纤维、耐酸玻璃纤维、普通玻璃纤

GF and electrical conductive GF, etc.

维（无碱玻璃纤维和中碱玻璃纤维）、光学玻璃纤维、低介电常数玻璃纤维和导电玻璃纤维等。

2 Preparation/ 制备

According to requirement for different GF, silica sand, quartz, **boric acid, clay** and other raw materials are mixed in different proportions, melted in high temperature furnace, and spun from spinneret. GF is rapidly winded while cooling and forming. This method is similar to melt spinning of chemical fibers.

Two methods are commonly used in GF production: direct drawing in tank furnace and ball drawing in crucible. For direct drawing in tank furnace, mineral raw materials are ground, fed to unit **kiln**, melted by heavy oil combustion and drawn. It has the characteristics of large output, stable quality and low energy consumption. The leak plate with 800~4000 holes is used for drawing in tank kiln, and the diameter of monofilament is more than 11 microns. For ball drawing in crucible, the glass ball is melted by electric heating and then drawn, in which energy consumption is high and the quality is unstable. Crucible is mainly pottery crucible, platinum crucible and platinum-substitute crucible. Pottery crucible can only be used to produce high-alkali GF from flat broken glass. All-platinum crucible can resist high temperature and is applied to produce clean and pure GF. But a single furnace needs 3~4 kg platinum rhodium alloy, which is expensive. Now the platinum-substitute crucible is mainly used, in which the melting part is high temperature resistant pottery material, and platinum **rhodium** alloy is used for the drawing leak plate. Only 0.6 kg precious metal is needed in a single furnace, which saves cost. But its quality is not as good as that of

按照不同玻璃纤维的要求，将硅砂、石英、硼酸及黏土等原料按不同比例混合，在高温炉中熔融，然后从喷丝板喷出纺丝，在冷却成形的同时，快速卷绕得到玻璃纤维。这种方法与化学纤维的熔融纺丝类似。

生产玻璃纤维常用池窑法直接拉丝和球法坩锅拉丝两种方法。池窑法直接拉丝是将矿物原料磨细、送入单元窑，用重油燃烧加热熔化后拉丝，具有产量大、质量稳定、能耗低的特点。池窑拉丝用漏板为800~4000孔，单丝直径在11 μm以上。球法坩锅拉丝是将玻璃球通过电加热熔化后拉丝，此法能耗大、质量不稳定。坩锅主要有陶土坩锅、全铂坩锅和代铂坩锅三种。陶土坩埚只能用于平板碎玻璃生产高碱玻璃纤维，全铂坩锅耐高温，能制备干净纯净玻璃纤维，但单炉需铂铑合金3~4 kg，造价昂贵。现在主要用代铂坩锅，熔化部分为耐高温陶土材料，拉丝漏板用铂铑合金，单炉只需贵金属0.6 kg，节省造价成本，但质量不如全铂坩锅。球法坩锅拉丝所用漏板为50~800孔，单丝直径

platinum crucible. The leak plate with 50~800 holes is used for drawing in crucible, and the diameter of monofilament is below 9 microns.

Surface treatment of GF is carried out with **sizing agent** and **coupling agent**. GF is treated by sizing agent to have lubrication protection and cohesive clustering effects, prevent the accumulation of electrostatic charges on GF surface and improve the **compatibility** with the matrix and the properties of chemical adsorption and bonding on the interface. Coupling agent treatment of GF has the functions of coupling, protection and property improvement of interface and composite. Coupling agent consists of two groups, in which one is bound to GF, and the other is bound to the matrix by physical or chemical action, which improves the GF surface adhesion. GF surface is protected by preventing from the intrusion of water or other harmful media. The interface weakness is reduced or eliminated to improve the interface state, which makes the stress transfer effectively. The mechanical properties, water resistance, chemical corrosion resistance, aging resistance, heat resistance and service life of GF reinforced composite are improved.

在 9 μm 以下。

采用浆料（浸润剂）和偶联剂对玻璃纤维进行表面处理。浸润剂处理玻璃纤维具有润滑保护和黏结集束作用，防止玻璃纤维表面静电荷的积聚，改善与基体的相容性及化学吸附和界面结合等性能。偶联剂处理玻璃纤维具有偶联、保护、改善界面性能和制作复合材料等作用。偶联剂含两种基团，一种基团与玻璃纤维结合，另一种基团与基体通过物理或化学作用相结合，改善玻璃纤维表面的黏结性。保护玻璃纤维表面，防止水分或其他有害介质的侵入。减少或消除界面弱点，改善界面状态，使应力有效传递。提高玻璃纤维增强复合材料的力学性能、防水性、耐化学腐蚀性、耐老化和耐热等性能及使用寿命。

3 Structure/ 结构

GF is mainly composed of silicon dioxide (SiO_2), diboron trioxide (B_2O_3), calcium oxide (CaO), aluminum oxide (Al_2O_3), etc., which have a decisive influence on the performance and manufacturing process of GF. **Silicate** GF and **borate** GF refer to the ones mainly composed of SiO_2 and B_2O_3, respectively. Alkaline oxide including **sodium oxide** (Na_2O), **potassium** oxide (K_2O), etc., acts as **fluxing agent** to destroy glass skeleton and loosen the structure. The melting temperature and viscosity of glass are reduced,

玻璃纤维的主要成分是二氧化硅（SiO_2）、三氧化二硼（B_2O_3）、氧化钙（CaO）和三氧化二铝（Al_2O_3）等，对玻璃纤维的性能和生产工艺起决定性作用。以二氧化硅为主的称为硅酸盐玻璃纤维，以三氧化二硼为主的称为硼酸盐玻璃纤维。氧化钠（Na_2O）和氧化钾（K_2O）等碱性

and the bubbles could be easily expelled from the glass solution. However, the higher the alkaline oxide content is, the lower the tenacity, electrical insulation and chemical stability of GF are. Therefore GF composition should not only mechanical propertiy and chemical stability, but also meet the requirements of manufacturing process.

氧化物作为助熔剂，破坏玻璃骨架，使结构疏松，降低玻璃的熔化温度和黏度，排除玻璃溶液中的气泡。但碱性氧化物的含量越高，玻璃纤维的强度、电绝缘性和化学稳定会降低。因此，玻璃纤维的成分既要满足力学性能和化学稳定性，又要满足制造工艺要求。

GF structure is shown in Figure 9-1. It is composed of SiO_2 **tetrahedral**, aluminum-oxygen **trihedral** or **boron-oxygen** trihedral to form irregular three-dimensional network. The gap between networks is filled with **cations** such as $Na^+, K^+, Ca^{2+}, Mg^{2+}$ and so on. The three-dimensional network structure of SiO_2 tetrahedral is the basis of determining GF property. The filled Na^+, Ca^{2+} and other cations are called network modifiers.

玻璃纤维结构如图9-1所示。其是由二氧化硅的四面体、铝氧三面体或硼氧三面体相互连成不规则三维网络，网络间的空隙由 Na^+、K^+、Ca^{2+}、Mg^{2+} 等阳离子所填充。二氧化硅四面体的三维网状结构是决定玻璃纤维性能的基础，填充的 Na^+、Ca^{2+} 等阳离子称为网络改性物。

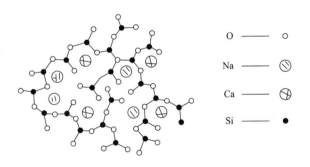

Figure 9-1 Diagram of GF structure
图9-1 玻璃纤维结构示意图

4 Property/ 性能

4.1 Appearance and Density/ 外观和密度

The cross section of GF is round. The smooth surface is not conducive to the cohesion among fibers and adhesion

玻璃纤维横截面为圆形，表面光滑，不利于纤维之间的抱合

between fiber and resin. The density of organic fiber and GFs are shown in Table 9-2. The density of GF is generally 2.16~4.30 g/cm^3, higher than that of organic fiber, lower than that of metal, and equal to that of aluminum. In addition, alkali-free GF has higher density than alkali GF.

及纤维与树脂的黏结。有机纤维和玻璃纤维的密度见表9-2。玻璃纤维密度一般在2.16~4.30 g/cm^3，大于有机纤维，小于金属，与铝密度相当。此外，无碱玻璃纤维密度比有碱玻璃纤维的大。

Table 9-2　Density of organic fiber and GFs
表 9-2　有机纤维和玻璃纤维的密度

Fiber 纤维	Wool 羊毛	Silk 蚕丝	Cotton 棉	Rayon 人造丝	Nylon 锦纶	CF 碳纤维	GF 玻璃纤维	
							E 无碱	A 有碱
Density 密度（g/cm^3）	1.28~1.33	1.30~1.45	1.5~1.6	1.5~1.6	1.14	1.4	2.6~2.7	2.4~2.6

4.2　Mechanical Property/ 力学性能

The outstanding mechanical characteristic of GF is its high tensile tenacity. The tenacity of fiber and metal is shown in Table 9-3. The tenacity of GF is much higher than that of natural fiber, tens of times that of aluminum and lump glass of the same composition. The factors affecting GF tenacity include diameter and length, chemical composition, storage time and loading time. The smaller the diameter, the higher the tenacity (Table 9-4). The tenacity decreases significantly with the increase of length(Table 9-5). The higher the alkali content, the lower the tenacity. The tenacity of alkali-free GF is 20% higher than that of alkali GF (Table 9-6). Alkali-free GF has high forming temperature, fast hardening speed and large structural bond energy. After storage for a period of time, GF is eroded by moisture and oxygen in air, which results in fiber ageing and tenacity descent. GF tenacity decreases with the increase of loading time. When the ambient humidity is high, the moisture adsorbed in microcracks makes it expand rapidly,

玻璃纤维的最大力学特征是拉伸强度高。纤维和金属的强度见表9-3。玻璃纤维强度远远高于天然纤维，是铝和同成分块状玻璃的几十倍。影响玻璃纤维强度的因素：直径和长度、化学组成、存放时间和施加负荷时间。直径越小，强度越高（表9-4）。随着长度增加，强度显著下降（表9-5）。含碱量越高，强度越低，无碱玻璃纤维比有碱玻璃纤维的强度高20%（表9-6），无碱玻璃纤维成型温度高、硬化速度快、结构键能大。存放一段时间后，玻璃纤维受到空气中的水分和氧气侵蚀而老化，强度降低。玻璃纤维强度随着施加负荷时间的延长而降低，环境湿度较

which leads to a more significant decrease in tenacity under the action of external force.

高时，在外力作用下，吸附在微裂纹中的水分使微裂纹扩展迅速，强度下降更明显。

Table 9-3 Tenacity of fiber and metal
表 9-3 纤维和金属的强度

Fiber 纤维	Wool 羊毛	Flax 亚麻	Cotton 棉	Silk 蚕丝	Al 铝	Glass 玻璃	GF 玻璃纤维
Diameter 直径（μm）	~15	16~50	10~20	18	Lump块状	Lump块状	5~8
Tenacity 强度（MPa）	100~300	350	300~700	440	40~460	20~120	1000~3000

Table 9-4 Relation between tenacity and diameter of GF
表 9-4 玻璃纤维强度与直径的关系

Diameter 直径（μm）	4	5	7	9	11
Tenacity 强度（MPa）	3000~3800	2400~2900	1750~2150	1250~1700	1050~1250

Table 9-5 Relation between tenacity and length of GF
表 9-5 玻璃纤维强度与长度的关系

Length 长度（mm）	Diameter 直径（μm）	Tenacity 强度（MPa）
5	13	1500
20	12.5	1210
90	12.7	360
1560	13	72

Table 9-6 Tenacity of alkali-free and alkali GF
表 9-6 无碱和有碱玻璃纤维强度

Fiber 纤维	Diameter 直径（μm）	Tenacity 强度（MPa）
E/无碱	5.01	2000
A/有碱	4.70	1600

The **stress-strain curve** of GF is basically a straight line and presents brittleness without plastic deformation. Elastic modulus and elongation ratio of fiber and metal are

玻璃纤维的应力应变曲线基本是一条直线，没有塑性变形，呈现脆性特征。纤维和金属的弹

131

shown in Table 9-7. The elastic modulus of GF is not high, and the abrasion resistance and flexural resistance of GF are poor. Especially in wet environment, the moisture adsorbed on the surface of GF can accelerate the propagation of microcrack. GF is treated with 0.2% cationic **surfactant** to improve its **flexibility**, which results in the improvement of abrasion resistance by about 200 times.

性模量和伸长率见表9-7。玻璃纤维的弹性模量不高，玻璃纤维的耐磨性和耐折性很差，尤其在潮湿环境下，玻璃纤维表面吸附水分后能加速微裂纹的扩展，采用0.2%阳离子表面活性剂处理玻璃纤维可以改善其韧性，耐磨性提高约200倍。

Table 9-7　Elastic modulus and elongation ratio of fiber and metal
表9-7　纤维和金属的弹性模量和伸长率

Fiber 纤维	Elastic modulus 弹性模量（MPa）	Elongation ratio 伸长率（%）
E 无碱玻璃纤维	72000	3.0
A 有碱玻璃纤维	66000	2.7
Cotton 棉	10000~12000	7.8
Wool 羊毛	6000	25~35
Flax 亚麻	30000~50000	2~3
Aramid 芳纶	3000	20~25
High-alloy steel 高合金钢	160000	—
Aluminum alloy 铝合金	42000~46000	—

4.3　Thermal Property/ 热学性能

The thermal conductivity coefficient of fiber and other substances is shown in Table 9-8. The larger clearance between GF leads to the smaller volume density. And the thermal conductivity coefficient of air is small. So GF has smaller thermal conductivity coefficient. GF is an amorphous polymer with glass transition temperature(Tg)

纤维和其他物质的导热系数见表9-8。玻璃纤维之间的空隙较大，容积密度较小，空气导热系数小，因此，玻璃纤维导热系数较小。玻璃纤维为无定型高聚物，存在玻璃化转变

and viscous flow temperature. The softening temperature ranges from 550 ℃ to 580 ℃ and T_g is about 600 ℃. GF does not burn and has good thermal resistance. GF tenacity remains unchanged below 200~250 ℃. Its thermal expansion coefficient is 4.8×10^{-6}/℃.

温度（T_g）和黏流温度。软化温度为550~580 ℃，T_g 约为600 ℃。玻璃纤维不燃烧，耐热性好，200~250 ℃以下，玻璃纤维强度不变。热膨胀系数为 4.8×10^{-6}/℃。

Table 9-8 Thermal conductivity coefficient of fiber and other substances
表9-8 纤维和其他物质导热系数

Fiber 纤维	Volume density 容积密度（kg/m³）	Thermal conductivity coefficient 导热系数[W/(m·K)]
Cotton 棉	81	0.058~0.062
Flax 亚麻	130	0.046~0.053
Silk 蚕丝	100	0.046~0.052
Wool 羊毛	80	0.034~0.046
GF 玻璃纤维	80	0.034
Glass 玻璃	—	0.7~1.3
Air 空气	—	0.0246
Water 水	—	0.6

4.4 Electrical Property/ 电学性能

The electrical property of GF is primarily determined by its chemical composition, temperature and humidity. The electrical insulation of alkali-free GF is better than that of alkali GF due to less alkali metal ions in alkali-free GF. The electrical resistivity of GF decreases with the increase of temperature. The addition of iron oxide, lead oxide, copper oxide, **bismuth** oxide and **vanadium** oxide in GF provides it with electrical property of semiconductor. Coating metal or graphite on GF allows it to be **conductive**.

玻璃纤维的电学性质主要取决于化学组成、温度和湿度。由于无碱玻璃纤维中碱金属离子少，其电绝缘性能好于有碱玻璃纤维。玻璃纤维的电阻率随着温度的升高而下降。玻璃纤维中加入氧化铁、氧化铅、氧化铜、氧化铋以及氧化钒，使其有半导体的电学性质。在玻璃纤维上涂覆金属或石墨，使其导电。

4.5 Chemical Property/ 化学性能

The chemical property of GF is significantly influenced by chemical composition, temperature, pressure and so on. Generally speaking, the chemical stability of GF mainly depends on the contents of silica and alkali metal oxide in its composition. Silica can improve the chemical stability of GF, while alkali metal oxide can reduce the chemical stability. Except hydrofluoric acid (HF), concentrated alkali (NaOH) and concentrated phosphoric acid (H_3PO_4), GF has good chemical stability for all chemicals and organic solvents.

Medi-alkali GF has better acid resistance than alkali-free GF (Table 9-9). Acid segregates and dissolves metal oxides (Na_2O, K_2O) on GF surface. Acid reacts with silicate in GF to form **silicic acid**, which rapidly polymerizes and **coagulates** into colloid, forming a very thin silica protective film on GF surface. Na_2O and K_2O are beneficial to the formation of the protective film.

The water resistance of alkali-free GF is better than that of medium-alkali GF. Firstly, water erodes alkali metal oxide on GF surface. Because water presents alkaline, then the reaction between GF and alkali solution continues over time until the silica skeleton is destroyed.

化学组成、温度和压力等对玻璃纤维的化学性质有重要影响。一般来说，玻璃纤维的化学稳定性主要取决于其成分中二氧化硅及碱金属氧化物的含量。二氧化硅有助于提高玻璃纤维化学稳定性，而碱金属氧化会使化学稳定性降低。除氢氟酸（HF）、浓碱（NaOH）、浓磷酸（H_3PO_4）外，玻璃纤维对所有化学药品和有机溶剂有很好的化学稳定性。

中碱玻璃纤维耐酸性好于无碱玻璃纤维（表9-9）。酸使玻璃纤维表面的金属氧化物（Na_2O、K_2O）离析和溶解。酸与玻璃纤维中硅酸盐反应生成硅酸，硅酸迅速聚合并凝成胶体，在玻璃纤维表面形成一层极薄的氧化硅保护膜，Na_2O 和 K_2O 有利于这层保护膜的形成。

无碱玻璃纤维的耐水性好于中碱玻璃纤维。首先，水侵蚀玻璃纤维表面的碱金属氧化物，由于水呈现碱性，然后随着时间增加，玻璃纤维与碱液继续反应，直至使二氧化硅骨架破坏。

Table 9-9 Performance comparison of alkali-free and medium-alkali GFs
表 9-9 无碱和中碱玻璃纤维性能对比

Fiber 纤维	Acid resistance 耐酸性	Water resistance 耐水性	Tenacity 强度	Anti-aging 防老化	Electrical insulation 电绝缘性	Cost 成本	Application condition 适用条件
E 无碱	Common 一般	Good 好	High 高	Better 较好	Good 好	High 高	High tenacity 强度高
C 中碱	Good 好	Bad 差	Lower 较低	Worse 较差	Low 低	Low 低	Low tenacity 强度高

5 Application/ 应用

GF has provides many desirable properties including high tensile tenacity, high temperature endurance, non-flammability, corrosion resistance, thermal insulation, sound insulation and electrical insulation, etc. GF is used in a number of applications, which can be divided into four basic categories: insulation, filtration media, reinforcement and optical fiber. GF is usually used as reinforcing material in composite, electrical insulation material, thermal insulation material and circuit substrate, etc.

(1) Textiles

GF can be processed into decorative textiles. It is mildew-proof, rot-proof, resistant to ageing, ultraviolet-proof, dimensionally stable, thermal insulating, non-staining, easy to maintain.

GF as warp yarn and staple yarn as weft yarn are woven into interwoven fabric, which can be used for wall cloth. It is suitable for the decorative material for **gypsum** wall surface due to **anti-skidding** and high tensile tenacity. Fabric made from GF filament is suitable for **asphalt** felt, roof felt, and cushion and reinforcement of special pavement. GF fabric can be made into high temperature proof or fire-proof clothing. Metalized GF fabric is obtained by coating metal membrane on it in a vacuum.

(2) Construction and Transportation Material

In addition to its use as textile materials, GF is extensively used as the thermal insulation material for residence, transportation and cold storage, warm retention and fire protection material for pipeline, sound absorption material for interior wall and ceiling in construction. And it also can be used in smelting furnace, dust removal

玻璃纤维具有拉伸强度高、耐高温、不燃、耐腐蚀、隔热、隔音和电绝缘等优点。玻璃纤维有多种用途，可分为四大基本类别：绝缘、过滤介质、增强体和光纤。玻璃纤维通常用作复合材料中的增强材料、电绝缘材料、绝热保温材料和电路基板等。

（1）纺织品

玻璃纤维不霉不烂、耐老化、防紫外线、尺寸稳定、隔热、不易沾污和方便维护，可用于装饰用纺织品。

玻璃纤维作为经纱与短纤纱作为纬纱的交织物可作为墙布。织物防滑、拉伸强度高，适合做石膏墙体表面的装潢材料。玻璃纤维长丝织物适用于沥青毛毡、屋顶毡及特殊路面的垫层和加固层。玻璃纤维布可制成防高温服或防火服。利用金属真空镀膜的方式包覆玻璃纤维布制作金属化玻璃纤维布。

（2）建筑交通材料

除了用作纺织材料外，玻璃纤维广泛用作住宅、交通工具和冷库的绝热材料、管道保温防火用材料、建筑如内壁、天花板屋顶的吸声材料，也可用于熔炉、发电厂和水泥厂的除尘设备中。

instrument for power and **cement** plants.

GF reinforced cement composite can improve the tensile tenacity, flexural tenacity and impact resistance of cement matrix. GFRP is a composite with light weight and high tenacity, which is made of thermosetting or thermoplastic resin as matrix and GF or glass cloth as reinforcement. It has good thermal resistance, corrosion resistance and electrical insulation.

（3）Electrical Insulation Material

GF can be processed into insulating impregnating product, GF reinforced plastic laminate, GF molding material and electromagnetic wire due to its excellent electrical insulation.

（4）Aerospace Material

With the vigorous development of aerospace field in China, high performance GF reinforced composite has become an indispensable material in aerospace industry. GF reinforced composite is used in civil and military.

玻璃纤维增强水泥复合材料可以提高水泥基的拉伸强度、弯曲强度和抗冲击性能。玻璃钢是一种以热固性或热塑性树脂为基体、玻璃纤维或玻璃布为增强体的轻质高强复合材料，耐热、耐腐蚀以及电绝缘性良好。

（3）电绝缘材料

由于较好的电绝缘性，玻璃纤维可加工成绝缘浸渍制品、玻璃纤维增强塑料层压制品、玻璃纤维模塑材料、电磁线。

（4）航空航天材料

随着我国航空航天领域的大力发展，高性能玻璃纤维增强复合材料已成为航空航天工业中不可或缺的一种材料，无论是民用还是军用都使用玻璃纤维增强复合材料。

Exercises/ 练习

1. What are the differences betweens glass fiber and common organic fibers？
2. Translate the following Chinese into English.

（1）相比金属纤维、天然纤维以及有机纤维，玻璃纤维强度高、密度小、吸湿低、延展性小、电绝缘性好、耐腐蚀、耐高温。

（2）氧化钠、氧化钾等碱氧化物作为助熔剂，通过破坏玻璃骨架，使得结构疏松，以降低玻璃熔化温度和黏度，并使玻璃中的气泡易排除。

（3）一般玻璃纤维的最大力学特征是抗拉强度高。玻璃纤维的应力—应变曲线基本是一条直线，没有塑性变形，呈现脆性特征。

（4）玻璃纤维的电学性质主要取决于化学组成、温度和湿度。无碱玻璃纤维电绝缘性能要好于有碱玻璃纤维，主要是由于无碱玻璃纤维中碱金属离子少。

（5）玻璃纤维增强水泥复合材料可以提高水泥基的拉伸强度、弯曲强度和抗冲击性能。玻璃钢是一种以热固性或热塑性树脂为基体、玻璃纤维或玻布为增强体的质轻高强的复合材料，耐热、耐腐蚀以及电绝缘性良好。

References/ 参考文献

［1］KUMRE A, RANA R S, PUROHIT R. A Review on mechanical property of sisal glass fiber reinforced polymer composites [J]. Materials Today: Proceedings ,2017, 4(2): 3466–3476.

［2］HEARLE J W S. High-performance fibers [M]. Cambridge: Woodhead Publishing Limited, 2001.

［3］晏雄. 产业用纺织品［M］. 上海：东华大学出版社，2010.

［4］张以河. 复合材料学［M］. 北京：化学工业出版社，2011.

Chapter Ten Aromatic Polyamide Fiber/ 芳香族聚酰胺纤维

1 Introduction/ 前言

The following designation was adopted in 1974 by the United States Federal Trade Commission to describe aromatic **polyamide** fiber under the generic term **aramid**: "a manufactured fiber in which fiber-forming substance is a long chain synthetic polyamide in which at least 85% of the **amide** (–CO–NH–) linkages are attached directly to two aromatic rings". Aromatic polyamide fiber includes poly(m-phthaloyl-m-phenylenediamine) (PMIA) fiber, namely Aramid 1313, and poly(p-phenylene terephthamide) (PPTA) fiber, namely aramid 1414.

As early as the 1960s, DuPont developed thermal and electrical insulating **meta-aramid** Nomex and **para-aramid** Kevlar with much higher tenacity and modulus (Nomex and Kevlar are DuPont Registered Trademarks). PPTA fiber was manufactured in China in the early 1980. At the end of the 1980s, another para-aramid Twaron (Twaron is a registered product of Dijing Company) appeared on the market, similar to Kevlar, which is an aromatic **copolyamide**. It originated from rigid para aramid chains, with which more flexible fibers with high tenacity can be developed. The aromatic copolyamide fiber was pioneered by Ozawa and Matsuda. The Technora (Technora is a registered product of Teijin) fiber was commercialized by Teijin. In the 1970s, Monsanto developed an X50 aromatic

1974年，美国联邦贸易委员会采用以下定义来描述芳香族聚酰胺纤维，通用术语为芳纶，即一种人造纤维，成纤物质是长链合成聚酰胺，其中至少85%的酰胺键（–CO–NH–）直接连接到两个芳香环上。芳香族聚酰胺纤维包括聚间苯二甲酰间苯二胺（PMIA）纤维即芳纶1313和聚对苯二甲酰对苯二胺（PPTA）纤维即芳纶1414两种。

早在20世纪60年代，杜邦公司开发了热和电绝缘的间位芳纶Nomex和更高强度和模量的对位芳纶Kevlar（Nomex和Kevlar是杜邦公司注册的商标），我国于20世纪80年代初期生产PPTA纤维。20世纪80年代末，另一种对位芳纶Twaron（Twaron是帝京公司的注册产品）上市，类似于Kevlar，是一种芳香族共聚酰胺。芳香族共聚酰胺源于刚性的对位芳纶链，可开发更柔韧的高强纤维，Ozawa和Matsuda开创了芳香族共聚酰胺纤维。帝

copolyamide fiber of aromatic **polyamide hydrazine.**

京公司商业化了 Technora 纤维（Technora 是帝京公司的注册产品）。20 世纪 70 年代，Monsanto 开发了一种芳香族聚酰胺肼的 X50 芳香族共聚酰胺纤维。

2 Preparation/ 制备

2.1 Polymer synthesis/ 高聚物合成

Aramid is prepared by the reaction between an **amine** group and a **carboxylic** acid **halide** group. Simple AB homopolymers may be synthesized in the following way. AABB homopolymer can also be obtained by similar reaction method. AABB aromatic polyamide is prepared from aromatic **diamine** and **diacid** or dicarboxylic acid. It can be synthesized by interfacial polymerization, low temperature polycondensation and melt or gas phase polymerization.

由胺基和羧酸卤化物基团之间的反应制备芳纶，根据以下方式合成简单的 AB 均聚物。AABB 芳香族聚酰胺由芳香族二胺和二酸或二元酸制备而成。可采用界面聚合、低温缩聚和熔体或气相聚合等合成方式。

$$n\text{NH}_2-\text{Ar}-\text{COCl} \longrightarrow \underset{A}{}[\text{NH}-\text{Ar}-\underset{B}{\text{CO}}]_n + n\text{HCl}$$

According to Figure 10-1, **PPTA** is usually synthesized by **polycondensation** of *p*-**phenylene diamine** (PPD) and **terephthaloyl chloride** (TCL) at low temperature. The appropriate amount of PPD dissolved in mixture of **hexamethylphosphoramide** (HMPA) and ***N*-methylpyrrolidone** (NMP), which was cooled in an ice/acetone bath to 258K (-15 ℃) in nitrogen. Then TCL was added with rapid **stir**ring to obtain thick, paste-like gel. Stop stirring, and the reaction mixture stayed overnight and gradually rose to room temperature. Finally, the **reactant mixture** was cleaned with water to remove the solvent and **hydrochloric acid** (HCl), and the polymer was collected by filtering. The **stoichiometry** of the solvent and **reactant**

根据图 10-1，PPTA 通常是由对苯二胺（PPD）和对苯二甲酰氯（TCL）经低温缩聚反应合成的。适量 PPD 溶解在六甲基磷酰胺（HMPA）和 *N*- 甲基吡咯烷酮（NMP）混合物中，在氮气中，在冰／丙酮浴中冷却至 258 K（-15 ℃），然后边快速搅拌边添加 TCL，得到黏稠的膏状凝胶。停止搅拌，反应混合物静置过夜，逐渐升至室温。最后用水清洗反应混合物，除去溶剂和盐酸（HCl），过滤收集聚

mixture is important for determining the **molar mass** of the final product. For example, Bair and Morgan reported that PPTA product with highest **inherent viscosity** was prepared when the volume ratio of HMPA: NMP was 2:1. The optimum concentration of reactant PPD and TCL was about 0.25 mol/L. When reactant concentration was less than 0.25 mol/L, the inherent viscosity decreased quite rapidly. When the reactant concentration was more than 0.3 mol/L, the inherent viscosity decreases slowly.

合物。溶剂和反应混合物的化学计量对确定最终产物的摩尔质量很重要。例如，Bair 和 Morgan 报道了 HMPA:NMP 的体积比为 2:1 时，制备出最高特性黏度的 PPTA，反应物 PPD 和 TCL 最佳浓度约为 0.25 mol/L。反应物浓度小于 0.25 mol/L，特性黏度迅速降低。反应物浓度大于 0.3 mol/L，特性黏度下降较缓慢。

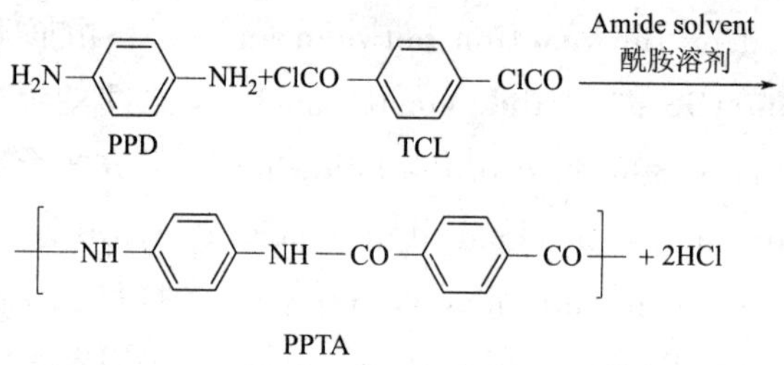

$$H_2N-\underset{PPD}{\underset{|}{\bigcirc}}-NH_2 + ClCO-\underset{TCL}{\underset{|}{\bigcirc}}-ClCO \xrightarrow{\text{Amide solvent / 酰胺溶剂}}$$

$$\left[-NH-\underset{}{\underset{|}{\bigcirc}}-NH-CO-\underset{}{\underset{|}{\bigcirc}}-CO-\right]_{PPTA} + 2HCl$$

Figure 10-1 PPTA synthesized by low temperature polycondensation of *p*-phenylene diamine (PPD) and terephthaloyl chloride (TCL)

图10-1 对苯二胺（PPD）和对苯二甲酰氯（TCL）低温缩聚合成PPTA

The reason for the drop-off of inherent viscosity at low reactant concentration is the solution polymerization of **aromatic diamine** and **aromatic dicarboxylic acid chloride** in ***N, N*-dimethyl acetamid**e (DMAc). The drop-off in inherent viscosity at the higher reactant concentration can be ascribed to decrease in reactant **mobility** due to **gelation** before obtaining high inherent viscosity. In addition, because the polymerization is **exothermic** and the higher reactant concentration leads to the generation of more heat, resulting in an increase in side reaction rate and a decrease in inherent viscosity.

Solvent mixture/salt system was utilized by Higashi et al. As shown in Figure 10-2, the polycondensation of **terephthalic acid** (TPA) and PPD was carried out to form high molecular weight PPTA in NMP containing **calcium**

低反应物浓度的特性黏度下降的原因是芳香二胺和芳香二羧酸氯化物在 *N, N-*二甲基乙酰胺（DMAc）中发生了溶液聚合。高反应物浓度的特性黏度下降的原因是在获得高特性黏度之前，由于凝胶化反应物流动性降低。另外，由于聚合反应是放热的，反应物浓度高，会产生更多的热量，导致副反应速率增加，降低特性黏度。

Higashi 等利用溶剂混合物/盐体系，如图 10-2 所示，对苯二甲酸（TPA）和 PPD 在含有氯化钙（CaCl$_2$）和氯化锂

chloride ($CaCl_2$) and **lithium chloride** (LiCl) in the presence of **pyridine**. The direct polycondensation reaction of aromatic **dicarboxylic** acid and **diamine** using **diphenyl** and **trialkyl phosphate** in NMP/pyridine solvent mixtures containing LiCl was discussed by Higashi et al.

（LiCl）的NMP中，在吡啶存在下进行缩聚形成高分子量PPTA。Higashi等探讨了利用二苯和三烷基磷酸酯在含氯化锂的NMP/吡啶溶剂混合物中的芳香族二羧酸和二胺的直接缩聚反应。

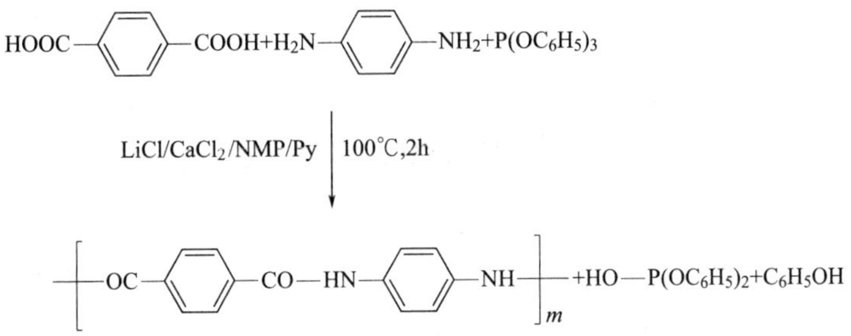

Figure 10-2 Polycondensation of terephthalic acid and *p*-phenylene diamine.
图10-2 对苯二甲酸与对苯二胺的缩聚

2.2 Spinning/纺丝

The polymer dissolves in **sulphuric acid** (H_2SO_4) at 80 ℃ to form liquid crystal solution. Polymer spinning solution is extruded through spinneret holes and enters the air coagulation bath. The velocity of the fiber as it leaves the **coagulating bath** is higher than the velocity of the polymer as it emerges from the spinning holes. Therefore the stretching takes place. This ratio is often referred to as the draw ratio. The drawing improves the orientation of liquid crystal region and higher mechanical properties of fiber are achieved. The fiber tenacity is affected by molecular orientation, the drying condition, temperature and tension.

Ozawa and Matsuda referred to three modes of drawing. An ordinary drawing mechanism would prevail in the case of low drawing ratio. An intermediate drawing mechanism for a medium drawing ratio (2~14) is dominated by shearing. An excessive drawing mechanism (higher than 15), is related to the formation of **microvoid**s and **microfibril**s.

聚合物纺丝液通过喷丝孔挤出，进入空气凝固浴，由于纤维离开凝固浴时的速度高于聚合物从喷丝孔中出来时的速度，从而发生牵伸，这个速度之比通常被称为牵伸倍数。牵伸提高液晶区取向，获得较高的纤维力学性能。纤维强度受分子取向、干燥条件、温度和张力影响。

Ozawa和Matsuda提出了三种牵伸模式。低牵伸倍数时，一般牵伸机理占主导。中等牵伸倍数（2~14倍）的牵伸机理以剪切为主。过度牵伸机理（高于15倍）与微孔隙和微原纤的形

After heat treatment for several seconds under tension, the orientation and modulus of fiber increase. An orientation angle of 12°~15° decreases to about 9° and the modulus increases from 64 GPa to over 150 GPa. The heat treatment effect of aramid is affected by molecular orientation, structure and spinning method.

张力下热处理数秒后，纤维取向和模量增加，取向度12°~15℃降低到约9°，模量从64 GPa增加到150 GPa以上。分子取向、结构和纺丝方法都会影响芳纶热处理效果。

3 Structure/ 结构

Molecular structure determines its physical properties. Dobb and Mcintyre has pointed out that the fiber tensile modulus depends largely on the molecular orientation along the fibre axis and the effective cross-sectional area of the single chain. In PPTA polymer chains are very rigid due to the para-connection of rigid benzene rings. The existence of regular amide groups along the main chain of linear macromolecules promotes a large number of transverse hydrogen bonding between adjacent chains, leading to effective chain stacking and high crystallization. On the contrary, in Nomex, benzene and amide units are adjacent to each other, forming irregular chain conformation and lower tensile modulus.

Northolt, Haraguchi and Yabuki et al. reported the PPTA structure. Northolt and Chaoy reviewed the relationship between the microstructure and mechanical properties of aramid. They found two kinds of crystal modifications of PPTA and proposed a model of PPTA crystal and molecular structure. It is considered that the chain conformation is mainly controlled by the interaction between **conjugated groups** in the chain. These interactions arise from the resonance effect due to the attempt to stabilize the **coplanarity** of amide and phenylene groups. **A counteracting steric hindrance** is

分子结构决定其物理性质。Dobb 和 Mcintyre 指出纤维拉伸模量在很大程度上取决于沿纤维轴的分子取向和单链所占的有效截面面积。在 PPTA 中，聚合物链由刚性苯环在对位连接而产生，非常刚硬，沿着线性大分子主链有规律的酰胺基的存在促进相邻链间横向的大量氢键连接，导致链有效堆砌和高结晶。相反，在 Nomex 中，苯和酰胺单元在邻位连接，形成不规整链构象和较低拉伸模量。

Northolt、Haraguchi 和 Yabuki 等报道了 PPTA 的结构，Northolt 和 Chapoy 综述了芳纶的微观结构与力学性能之间的关系。他们发现 PPTA 的两种晶体修饰，提出一种 PPTA 晶体和分子结构的模型。认为链构象主要由链中共轭基团之间相互作用控制，这些相互作用产生于共振效应，是由于设法稳定酰胺基和亚苯基的共面性。在对苯二胺段的

also found between the oxygen and the ortho-hydrogen of the *p*-phenylenen diamine segment and between the amide hydrogen and the ortho-hydrogen of the terephthalic segment.

Dobb and his colleagues studied the supramolecular structure of high modulus aramid fibers (PPTA) by combining electron diffraction and electron microscopy dark field image. It has been found that PPTA is a regular folded chain structure, and the alternating components of each segment are arranged at approximately the same but opposite angles to the cross-section plane. The angle between the adjacent parts of the pleat is about 170°.

氧和邻氢之间以及对苯二甲酸段的酰胺氢和邻氢之间也出现了抵消空间位阻。

Dobb与团队采用了电子衍射和电子显微镜暗场成像技术相结合的方法研究了高模量芳纶（PPTA）的超分子结构，发现PPTA为规整的折叠链结构，其每个片段的交替组分与截面平面呈大致相等但相反的角度排列，折叠的相邻部分之间的夹角约为170°。

4 Property/ 性能

Aramid has unique properties different from other fibers. It has low fiber elongation and higher tenacity and modulus than early organic fibers. It also has resistance to organic solvent, fuel, lubricant and flame. Aramid is easier to be woven on fabric loom than brittle fibers such as glass fiber, carbon fiber or ceramic fiber.

芳纶具有不同于其他纤维的独特性能。纤维伸长较低，拉伸强度和模量明显高于早期的有机纤维。它还具有耐机溶剂、燃料、润滑剂和火焰性能。芳纶比玻璃纤维、碳纤维和陶瓷纤维等脆性纤维更易于在布机上织造。

4.1 Mechanical property/ 力学性能

By weight, aramid is stronger than steel wire and harder than glass. The tenacity and modulus of Kelvar are more twice and nine times those of high strength Nylon, respectively. Creep and linear thermal expansion coefficient are low and thermal stability is high, similar to inorganic fibers. All these are related to high molecular weight, straight chain and good orientation. Aramid has excellent properties related to organic fibers, such as low density, easy **processability** and fairly good fatigue and wear resistance.

按重量计算，芳纶比钢丝强，比玻璃硬。Kelvar拉伸强度和模量分别是高强锦纶的2倍多和9倍多，蠕变和线性热膨胀系数较低，热稳定性高，类似于无机纤维，这些都跟高分子量、伸直链和良好取向有关。芳纶具有有机纤维相关的优良特性，比如，低密度、易加工和相当好的

Therefore, aramid is practical.

Due to the flexibility of copolymer chain and the loose crystal structure in copolymer, Technora exhibits high fatigue resistance, and has comparable tensile property and thermal stability to those of highly crystalline para-aramid.

Creep is affected by temperature, fiber relative ultimate tenacity, water content and other parameters. The creep of para-aramid is very small, which is significantly different from that of other highly oriented polymer fibers. There are differences between Kevlar 49 and Kevlar 29. If the load is less than 50% of the fiber breaking tenacity, the creep amplitude of Kevlar 29 has no correlation with the changes of temperature and load.

Compression and shear properties are correlated with axial orientation and radial intermolecular bonding. The compressive tenacity of Twaron is 0.6 GPa, about one-fifth of its tensile tenacity. The shear modulus of para-aramid is higher than that of traditional man-made fibers and lower than that of its compression modulus, which is due to the anisotropy of its radial structure.

耐疲劳和耐磨性，因此，芳纶具有实用性。

由于共聚物链的柔韧性和共聚物中疏松的晶体结构，Technora 具有较高的抗疲劳性，Technora 拉伸性和热稳定性与高结晶对位芳纶相当。

蠕变受温度、相对纤维极限强度的负荷、含水量和其他参数的影响，对位芳纶蠕变很小，与其他高取向聚合物纤维有显著差异。Kevlar 49 和 Kevlar 29 之间存在差异。如果负荷低于纤维断裂强度的 50% 时，Kevlar 29 蠕变振幅与温度和负荷的变化没有相关性。

压缩和剪切性能与轴向取向和径向分子间结合相关。Twaron 压缩强度为 0.6 GPa，约为拉伸强度的五分之一。对位芳纶剪切模量高于传统再生纤维，低于其压缩模量，这是由于其径向结构的各向异性。

4.2 Thermal property/ 热学性能

Aramid has good thermal stability and high limited oxygen index (LOI) in the range of 27%~43%. It performs well when exposed to high temperature as high as 180 ℃ for a long period of time. It does not **embrittle** or degrade at the low temperature of −60 ℃. Its thermal expansion coefficient is very small and anisotropic. The longitudinal and transverse thermal expansion coefficients are -4×10^{-6}~-2×10^{-6}/ ℃ and 5.9×10^{-6}/ ℃, respectively.

Like other polymers, aromatic polyamide is sensitive

芳纶具有良好的热稳定性，高的极限氧指数（27%~43%），180 ℃能长期使用，在低温 −60 ℃不脆化、不降解。热膨胀系数很小，各向异性，纵横向热膨胀系数分别为 -4×10^{-6} ~ -2×10^{-6}/ ℃和 59×10^{-6}/ ℃。

与其他聚合物一样，芳香族

to radiation, especially in the range between about 300 nm and 450 nm. For outdoor application proper protection from radiation is necessary to maintain good mechanical properties. Aramid is susceptible to be eroded by acid and alkali. Especially it is not resistant to strong acid. It is also poor in water resistance due to the existence of polar **acylamino**. Properties of commonly used reinforcing fibers are shown in Table 10-1.

聚酰胺对辐射很敏感,特别是在波长为300~450 nm。对于室外应用,有必要采取适当的防辐射保护,以保持良好的力学性能。芳纶易受到酸碱的侵蚀,尤其是不耐强酸。由于极性酰胺基的存在,耐水性不佳。常用增强纤维的性能见表10-1。

Table 10-1 Properties of commonly used reinforcing fibers
表 10-1 常用增强纤维的性能

Material 材料	Density 密度（g/m³）	Decomposing temperature 分解温度（℃）	Tenacity 强度（mN/tex）	Initial modulus 初始模量（N/tex）
Para-aramid standard 标准对位芳纶	1.44	550	2065	55
High-modulus Para-aramid 高模量对位芳纶	1.45	550	2090	77
Nomex	1.46	415	485	7.5
Technora	1.39	500	2200	50.3
PA 66 锦纶66	1.14	255	830	5
Steel cord 钢丝绳	7.85	1600	330	20
Carbon HT 高强碳纤维	1.78	3700	1910	134
Carbon HM 高模碳纤维	1.83	3700	1230	256
E-glass E-玻璃纤维	2.58	825	780	28

5 Application/ 应用

（1）Para-aramid

There are vast applications for Para-aramid. It can be used to fabricate tire cord, especially suitable for truck and aircraft tyres. It is an ideal fiber reinforcing material, which can be used for structural materials of aircraft and

（1）对位芳纶

对位芳纶用途极为广泛,用于轮胎帘子线的制作,特别适合于载重汽车飞机的轮胎;是极为理想的纤维增强材料,用于飞机

spacecraft, rocket motor shells, and bullet-proof products such as helmet, vest and so on. It also has important applications in rope, gloves, sports goods and so on.

(2) Meta-aramid

It can be used as filtering materials for high temperature, chemicals, smoke and so on. It can also be used for fire-proof curtain, warm-retention clothing, battle uniform, blanket, thermal-endurance parachute, and high-temperature endurance conveyor belt. End-use, field and key properties of Aramid is shown in Table 10-2.

和宇航器的结构材料、火箭发动机壳体材料以及头盔、背心等防弹制品；在绳索、手套、体育用品等方面也有重要的应用。

（2）间位芳纶

用于高温、化学品、烟尘等过滤，可制作防火帘、隔热服、作战服、地毯、耐热降落伞等，以及工业用耐高温传输带。芳纶的用途、领域及其关键属性见表 10-2。

Table 10-2　End-use, field and key properties of Aramid
表 10-2　芳纶最终用途、领域及其关键属性

End-use 最终用途	Field 领域	Key properties 关键属性
Composites 复合材料	Fabrics for aircrafts & containers 飞机和集装箱用织物	Light weight, High tenacity, High modulus 重量轻，高强，高模 Good impact tenacity, Wear resistance 良好的冲击强度，耐磨
	Pressure vessels 压力容器	
	Ship building 造船	
	Sport goods 体育用品	
	Plastic additive 塑料添加物	
	Civil engineering 土木工程	
Protective apparels 防护服	Heat resistance workwear 耐热工作服	Heat resistance, Flame retardation, Cut resistance 耐热，阻燃，防割
	Fire blankets 阻燃电探	
	Flame retardant textiles 阻燃纺织品	
	Cut protective gloves 防割手套	
	Cut protective seat cover layers 防割座椅罩层	
	Heat resistance workwear 耐热工作服	

续表

End-use 最终用途	Field 领域	Key properties 关键属性
Types 轮胎	Truck and aircraft tyres 卡车和飞机轮胎	Low density, Weight saving, High tenacity 低密度，减轻重量，高强 Dimensional stability, Low shrinkage, Puncture resistance 尺寸稳定，低收缩，防刺
	High speed tyres 高速轮胎	
	Motorcycle tyres 摩托车轮胎	
	Bicycle tyres 自行车轮胎	
Mechanical rubber goods 机械橡胶制品	Conveyor belts 输送带	High tenacityn, High modulus, Dimensional stability 高强，高模，尺寸稳定 Thermal resistance, Chemical resistance 耐热性，耐化学性
	Transmission belts 传动带	
	Hoses for automotive 汽车用软管	
	Hydraulic hoses 液压软管	
	Hoses in off-shore 离岸软管	
	Umbilicals 管缆	
Friction products and gaskets 摩擦产品和垫圈	Brake linings 刹车片	Fiber fibrillation, Heat resistance, Chemical resistance 纤维原纤化，耐热性，耐化学性 Low flammability, Mechanical performance 低可燃性，力学性能
	Clutch facings 离合器面	
	Gaskets 垫圈	
	Industrial paper 工业用纸	
Ropes and cables 绳索和电缆	Aerial optical fiber cable 架空光缆	High tenacity, High modulus, Dimensional stability 高强，高模，尺寸稳定 Low density, Corrosion resistance 低密度，耐腐蚀 Good dielectric properties, Heat resistance 良好的介电性能，耐热
	Traditional optical fiber cable 传统光缆	
	Electrocable 电缆	
	Mechanical construction cable 机械施工电缆	
	Mooring ropes 系泊缆绳	

续表

End-use 最终用途	Field 领域	Key properties 关键属性
Life protection 生命防护	Bullet proof vests 防弹背心	High tenacity, High energy dissipation 高强,高耗能 Low density and weigh reduction, Comfort 低密度和重量减轻,舒适
	Helmets 头盔	
	Property protection panels 性能保护面板	
	Vehicle protection 车辆防护	
	Strategic equipment shielding 战略装备屏蔽	

Exercises/ 练习

1. What is the main property difference between of para-aramid fiber and meta-aramid fiber?

2. Translate the following Chinese into English

(1) 杂环芳族聚酰胺纤维是指含有氮、氧、硫等杂质原子的二胺和二酰氯缩聚而成的芳酰胺纤维,如有序结构的杂环聚酰胺纤维等。

(2) 芳纶质轻,相对密度仅为1.44~1.45,因此,芳纶具有高的比强度和高的比模量,可应用于要求高强、高模的场合和要求高耐冲击的场合,芳纶压缩强度不高,剪切强度也不高。

(3) 芳纶易受到酸碱的侵蚀,尤其是强酸抵抗力较弱。由于芳纶分子中存在极性基团酰胺基,使其耐水性不佳。

(4) 芳纶具有优良的耐磨性能,利用芳纶的优良耐磨性能,可以将其应用于汽车轮胎、刹车片等耐磨制品的制造。

(5) 芳纶用途极为广泛,用于轮胎帘子线的制作,特别适合于载重汽车、飞机的轮胎;是极为理想的纤维增强材料,用于飞机、宇航器的结构材料、火箭发动机壳体材料;其制作的防弹制品(头盔、背心等)防弹性能优良;在绳索、手套、体育用品等方面也有重要的应用。

References/ 参考文献

［1］HEARLE J W S. High-Performance Fibers [M].Cambridge: Woodhead Publishing Limited, 2001.
［2］晏雄. 产业用纺织品 [M]. 上海：东华大学出版社，2010.
［3］张以河. 复合材料学 [M]. 北京：化学工业出版社，2011.

Chapter Eleven Ultra-high Molecular Weight Polyethylene Fiber/ 超高分子量聚乙烯纤维

1 Introduction/ 前言

Gel-spun polyethylene (PE) fiber is ultra-strong, high modulus fiber that is based on the simple and flexible PE molecule. It is called high performance PE (HPPE) fiber, high modulus PE (HMPE) fiber or sometimes extended chain PE (ECPE) fiber.

Ultra high molecular weight polyethylene (UHMWPE) fiber, also known as high tenacity and high modulus PE fiber, is gel-spun based on PE with an average molecular weight of more than 10^6. It is the highest specific tenacity and modulus fiber in the world.

In the late 1970s, DSM Company in the Netherlands succeeded in spinning UHMWPE fiber by gel spinning, and began industrial production in 1990, with the **trademark** Dyneema. The company is the founder and manufacturer of UHMWPE fiber with highest production and best quality in the world, and with an annual output of about 5,000 tons. In the 1980s, Allied Singal Company of the United States purchased the patent of Dutch DSM Company, developed its own production process and industrialized it. In 1990, Allied Signal Company was merged by Honeywell Company and continued to produce UHMWPE fiber, under the

凝胶纺聚乙烯（PE）纤维是以简单、柔软的 PE 分子为基础的超强、高模量纤维，称为高性能聚乙烯（HPPE）纤维、高模量聚乙烯（HMPE）纤维，有时也称为伸直链聚乙烯（ECPE）纤维。

超高分子量聚乙烯纤维，又称高强高模聚乙烯纤维，是平均分子量在 10^6 以上的凝胶纺出的聚乙烯纤维，目前是世界上比强度和比模量最高的纤维。

20 世纪 70 年代末期，荷兰帝斯曼（DSM）公司采用凝胶纺丝方法纺制 UHMWPE 纤维获得成功，并于 1990 年开始工业化生产，商标为 Dyneema。该公司是 UHMWPE 纤维的创始公司，是世界上产量最高、质量最佳的制造商，年产量约 5000 吨。80 年代，美国 Allied Singal 公司购买了荷兰 DSM 公司的专利，开发出了自己的生产工艺并实现工

trade name Spectra, with an annual output of about 3000 tons. Toyobo Textile Company (Toyobo) in Japan and DSM Company in Holland jointly produce UHMWPE fiber in Japan, with the **brand** Dyneema. The sale area is limited to Japan and Taiwan province of China, with an annual output of about 600 tons.

China is a big chemical fiber country, but not a powerful chemical fiber country. According to experts and statistics of relevant departments, the output of high-performance fiber in China is only one percent of the output in the world. The three high performance fibers in the world today are aramid, carbon fiber and UHMWPE. At present, aramid fiber is only produced in small quantities in China due to technical problems. Carbon fiber is still in the experimental and primary production stage, and its products can only be used in the field of wear-resistant filler. Since Beijing Tongyizhong Special Fiber Technology Development Company Ltd. broke through the key production technology of UHMWPE fiber in 1994, many industrial **production base**s of UHMWPE fiber has been formed in China so far. It is reported that 70% of UHMWPE fiber in the United States is used in the military fields of bullet-proof armor, aerospace and so on, such as body armor, bullet-proof helmet, military facility and equipment. The development of high-performance fiber is an embodiment of the comprehensive strength of a country and an important material basis for building a modern powerful country. At present, because the development of the production and application of UHMWPE fiber have been strongly supported and accelerated in China, domestic UHMWPE fiber has occupied a pivotal position in the world.

业化。1990年，Allied Signal公司被霍尼韦尔（Honeywell）公司兼并，继续生产UHMWPE纤维，商标为Spectra，年产量约3000t。日本东洋纺（Toyobo）公司和荷兰DSM公司合资在日本生产UHMWPE纤维，商标为Dyneema，销售地区仅限日本和中国台湾，年产量约600t。

中国是化学纤维生产大国，但不是化学纤维生产强国，据专家介绍和有关部门统计，中国的高性能纤维的产量仅为世界产量的百分之一。当今世界三大高性能纤维是芳纶、碳纤维、UHMWPE纤维，当前中国由于技术问题芳纶仅有少量生产；碳纤维尚处在试验和初级生产阶段，产品只能应用在耐磨填料等领域；自1994年北京同益中特种纤维技术开发有限公司突破UHMWPE纤维的关键性生产技术以来，现已经在国内形成多个UHMWPE纤维产业化生产基地。据报道，美国超高分子量聚乙烯纤维70%用于防弹衣、防弹头盔、军用设施和设备的防弹装甲、航空航天等军事领域，而高性能纤维的发展是一个国家综合实力的体现，是建设现代化强国的重要物质基础。当前，由于国家大力支持和加速发展UHMWPE纤维的生产与应用，国产UHMWPE纤维已经在全世界占有举足轻重的地位。

2 Manufacturing/ 制备

PE is a flexible polymer with a very weak interaction between the molecular chains. This interaction is so weak that for strong fibers, ultra-long chains with a high overlap length are required. UHMWPE fiber is comprised of PE with very high molecular weight. The chemical properties of this PE are the same as those of HDPE, but its molecular weight is higher than that of the commonly used PE. The higher the molecular weight is, the higher the obtained tenacity is. Both the average molecular weight and the molecular weight distribution are critical parameters. Too long chains hinder the drawing and short chains are less effective in the load transmission of fiber. Quantitative branched chains give fibers better properties, but also interfere with drafting. PE molecules are much longer and more flexible. After physical treatment, the molecules form straight chain, high orientation and high crystallization in the fiber direction. In 1979 DSM invented and patented UHMWPE fiber and the gel-spinning process to produce it. Several further patents concerning this process have been filed in later years.

The UHMWPE gel spinning process includes the following steps: Firstly, UHMWPE dissolves in appropriate solvents to form a semi-dilute spinning solution. Secondly, the spinning solution is continuously extruded through spinneret orifices, **gelate**d and crystallized into UHMWPE as-spun fiber, which can be done either by sudden cooling in air or water and extraction, or by solvent evaporation. Thirdly, UHMWPE as-spun fiber is superdrawn into UHMWPE fiber and the remaining solvent is removed, which gives fiber final properties. The other steps are essential in the production of a fiber with good

PE 是一种柔性聚合物，其分子链之间的相互作用非常弱，形成强力纤维需要具有高搭接长度的超长链。UHMWPE 纤维是由非常高分子量 PE 制成，这种 PE 的化学性能与普通高密度聚乙烯（HDPE）相同，但分子量高于常用 PE，分子量越大，获得的强度就越高。平均分子量和分子量分布都是关键参数，太长链阻碍牵伸，短链的纤维荷载传递的效率较低，适量支链赋予纤维较好的性能，同时也会干扰牵伸。聚乙烯分子长而柔韧，经过物理处理，分子在纤维方向上形成直链、高取向和高结晶。1979年，DSM 发明了 UHMWPE 纤维及其生产方法——凝胶纺丝工艺并获得专利，在以后的几年里申请了其他几项与工艺相关的专利。

UHMWPE 凝胶纺丝工艺流程包括以下几个步骤：（1）UHMWPE 溶解在适当的溶剂中形成半稀纺丝溶液；（2）纺丝溶液经喷丝孔连续挤出，然后通过空气或水骤冷却萃取，或溶剂蒸发，凝胶结晶成 UHMWPE 初生纤维；（3）超牵伸 UHMWPE 初生纤维形成 UHMWPE 纤维，并去除剩余溶剂，赋予纤维最终性能。生产具有良好特性的纤维还

characteristics.

2.1 Spinning Solution/纺丝液

UHMWPE is very difficult to dissolve. It can dissolve only after stirring for a long time at 170 ℃, but the molecular weight of UHMWPE decreases sharply. UHMWPE with molecular weight than 10^6 dissolves in suitable solvent for gel spinning. The main solvents used for UHMWPE gel spinning incldue decalin, paraffin oil, paraffin and kerosene. UHMWPE can dissolve in decalin at lower temperature and the solution is uniform. The decalin is easy to volatilize, and the precursor can be drawn directly without extraction. The substitution of alkanes for decalin can reduce production cost, but the distillation range of alkanes is high and it is difficult to remove them during stretching, so the extraction process must be adopted. UHMWPE in solution is easily degraded when heated in air. Dissolution with thermal and oxygen insulation, and addition of antioxidants and thermal stabilizers in solution can prevent thermal oxidation degradation.

Due to the regular molecular chain, high crystallinity and molecular weight, UHMWPE is difficult to dissolve uniformly, and the macromolecular chain can not be fully disentangled. By uniform pre-swelling, the solvent infiltrates and diffuses into the polymer to the maximum extent, which weakens the strong interaction between macromolecular chains. The more sufficient the solvation is, the easier the solvent is to enter the dissolution stage, which leads to the homogeneous dissolution of UHMWPE. The UHMWPE swelling needs to be carried out at a specific temperature. The full solvation can not be obtained at low temperature. A high viscosity layer on the surface of the UHMWPE particles is formed at high temperature, which impedes the solvent penetration into their interior. It is also not sufficient

需其他步骤。

UHMWPE极难溶解，在170℃下长时间搅拌，才能溶解，但分子量急剧下降。将分子量大于10^6的UHMWPE溶解于适当的溶剂中，进行凝胶纺丝。用于UHMWPE凝胶纺丝的溶剂主要有十氢萘、石蜡油、石蜡和煤油，十氢萘可在较低温度下溶解UHMWPE，溶液均匀性好，十氢萘易于挥发，制得的原丝可不萃取而直接拉伸，以烷烃类取代十氢萘可降低生产成本，但烷烃类馏程高，在拉伸过程中难以去除，必须采用萃取工艺。溶液中UHMWPE在空气中受热容易降解，隔热绝氧溶解和在溶液中添加抗氧化剂及热稳定剂可阻止热氧化降解。

由于分子链规整、结晶度和分子量高，UHMWPE很难均匀溶解，大分子链不能充分的解缠。通过均匀预溶胀，溶剂最大限度地向聚合体内渗透和扩散，减弱大分子链之间的强相互作用力，这种溶剂化作用越充分，溶剂则越易进入溶解阶段，使UHMWPE均匀溶解。UHMWPE溶胀需在特定温度进行，温度低不能充分溶剂化，高温下，UHMWPE颗粒表面形成高黏度层，阻碍溶剂向其内部渗入，同样也不能充分溶剂化，还

to solvate but it is easy to form gel blocks. Therefore, the swelling temperature is the main parameter of the swelling process. After full swelling, UHMWPE molecules disperse freely into solvent, which still need high dissolution temperature. With the increase of the molecular weight of PE, there are still a number of instantaneous entanglement points in the swelled UHMWPE molecular chain. In order to disperse macromolecules into solution in the form of integral coils, these entanglement points must be removed at the same time. Therefore, increasing temperature enhances the UHMWPE salvation by the solvent. Different solvents have different solvation ability and dissolution temperature for UHMWPE. The dissolution temperatures of decalin, paraffin oil and kerosene for UHMWPE are 145~150 ℃, 170~190 ℃ and 180~250 ℃, respectively.

It is also important to control the solution concentration appropriately. When the solution concentration is lower, there is little or no entanglement between macromolecules and the slip between macromolecules is easy to occur during stretching, which is not conducive to the stretching of the whole super-long molecular chain. Extension at very low stretching rate can stretch super long molecular chains. This method has no practical significance and can not be industrialized. When the solution concentration is higher, the solution mobility is poor and the flow is unstable. There are too many entanglement points between macromolecules in the initial gel fiber. When the high stretching is carried out, the internal stress is concentrated on the tangled chain, and the high drawing can not be realized. The initial gel fiber is formed by the gel solution with suitable concentration. Under certain technological conditions, there are a number of entanglement molecules and entanglement network structures in the gel fiber, which form deformation. The molecular chain stretches along the tensile force direction, and the tensile force is transferred smoothly to

易出现凝胶块，因此，溶胀温度是溶胀工艺的主要参数。充分溶胀后，UHMWPE分子向溶剂中自由分散，仍需要高的溶解温度。随着PE分子量提高，溶胀后的UHMWPE分子链上仍有一定数量的瞬间缠结点。要使大分子以整体线团形式向溶液中分散，必须同时解除这些缠结点。因此，提高温度增强溶剂对UHMWPE的溶剂化作用。各种溶剂对UHMWPE的溶解作用能力不同，溶解温度也不同。十氢萘、石蜡油和煤油对UHMWPE的溶解温度依次为145~150 ℃、170~190 ℃和180~250 ℃。

适当地控制溶液浓度也很重要，溶液浓度较低时，大分子间的缠结很少或不存在，拉伸时大分子间很容易产生滑移，不利于整个超长分子链的伸展。以极低的拉伸速率进行拉伸，使超长分子链伸展，这种方法无实际意义，不可工业化。溶液浓度较高时，溶液流动性能差，流动不稳定，所得初生态凝胶纤维中大分子间缠结点太多，进行高倍拉伸时，内应力集中在缠结链上，无法实现高倍拉伸。只有适当浓度的凝胶溶液形成的初生态凝胶纤维，在一定的工艺条件，凝胶纤维中存在一定数量的缠结分子和缠结网络结构形成形变，分子链沿拉伸力方向伸展，拉伸力顺利传递，实现高倍拉伸。

achieve high drawing.

At present, when decalin and paraffin oil are used as solvents in industrialized UHMWPE fiber, the optimum solution concentrations are 15% and 8%, respectively. The solution is prepared by homogenizing the suspension solution, which then enters into the twin screw to prepare the continuous gel solution, and finally continuous extrusion, metering and spinning are done.

目前，已工业化生产的UHMWPE纤维以十氢萘和石蜡油为溶剂时，最佳溶液浓度分别为15%和8%。溶液的制备以悬浮溶液的均化配制，再进入双螺杆进行连续凝胶溶液的制备，继而进行连续挤出、计量和纺丝成型。

2.2 Gelation and Crystallization/ 凝胶化和结晶

The solvent used in the UHMWPE gel-spinning process should be a good solvent at high temperature (>100 ℃) but at lower temperatures (<80 ℃) the polymer should easily crystallize from the solution. At spinning temperature, the spinning solution is injected into the low temperature (or room temperature) gas through the spinneret and then cooled into a coagulation bath to form an initial gel fiber. Under shear of spinning solution in the spinneret channel, some of the solvent is precipitated, and a large amount of solvent retaines in the gel thread. In the 10~30 min for the gel filament generation, the movement of macromolecules is easier due to the plasticizing effect of the solvent. Therefore, a large amount of solvent is precipitated from the gel filament. At this time, the fiber drawing ratio is zero, but a small amount of crystal has been produced.

UHMWPE凝胶纺丝过程中使用是高温溶剂（>100 ℃），但在较低的温度（<80 ℃）时，聚合物应容易从溶液中结晶出来。在纺丝温度下，纺丝溶液通过喷丝孔后，进入低温（或室温）气体后，再进入凝固浴中冷却成形，形成初生态凝胶纤维。纺丝溶液在喷丝孔道内受剪切作用，部分溶剂被析出，大量的溶剂仍保留在凝胶丝条中，在凝胶丝生成过程的10~30 min内，由于溶剂的增塑作用，大分子运动比较容易，因此，大量溶剂会从凝胶丝中析出，此时，纤维拉伸倍数为零，但已产生少量结晶。

2.3 Extraction/ 萃取

The reagent with low boiling point and volatile reagent is selected as the extractant, and the gel fiber is formed by solvent extraction. Gasoline, hexane, xylene, carbon tetrachloride and chloroalkanes are generally used. If the extraction is insufficient and the solvent residue ration is

选用沸点较低且易挥发的试剂作为萃取剂，通过溶剂萃取形成凝胶纤维，一般选用汽油、正己烷、二甲苯、四氯化碳和氯代烷烃等。如果萃取不充分，溶剂

high, the fiber tenacity obtained by drawing can not meet the requirement. Therefore, the pre-draft before extraction can make the gel filament thinner, which is beneficial to speed up the exchange rate of solvent and extractant in the extraction process, and improve extraction efficiency.

残留率高，则经牵伸后得到的纤维强度不能达到要求。因此，在萃取前进行预牵伸，使凝胶丝条变细，有利于加快萃取过程中溶剂和萃取剂的交换速率，提高萃取效率。

2.4 Drawing/ 牵伸

Almost all the solvents are contained in the UHMWPE gel, so the entanglement state of the UHMWPE macromolecule chains is well maintained. Moreover, the decrease of solution temperature leads to the formation of UHMWPE **lamellar** crystals with folded chains in the gel. Through the thermal superstretching gel UHMWPE precursor, the macromolecular chain is fully oriented and highly crystallized, and the macromolecule with folded chain is transformed into the extended chain, thereby UHMWPE fiber with high tenacity and high modulus is prepared.

几乎所有的溶剂被包含在 UHMWPE 凝胶中，UHMWPE 大分子链的缠结状态被很好地保持下来，而且溶液温度的下降，导致凝胶中 UHMWPE 折叠链片晶的形成。通过热超牵伸凝胶 UHMWPE 原丝使大分子链充分取向和高度结晶，呈折叠链的大分子转变为伸直链，从而制得高强高模 UHMWPE 纤维。

Multistage thermal drafting with slower drafting speed is performed and gel filaments stay in the heat pipe for a certain time, so that they can be stretched for several dozen times. In general, the first-order drawing is carried out at a lower temperature (90 ℃) and the drawing ratio is larger, and the second-order drawing is carried out at a higher temperature (110 ℃) and the drawing ratio is smaller. The temperature of the hot drawing machine and the rotation speed of the front and rear guide rollers are adjusted to determine the hot drawing ratio, and the extracted dry gel threads are subjected to super stretching on the heat treatment machine.

进行多级热牵伸，牵伸速度较慢，丝条在热管中停留一定时间，这样才能使凝胶丝拉伸几十倍。一般一级拉伸在较低的温度（90 ℃）下进行，拉伸倍数较大；二级拉伸在较高的温度（110 ℃）下进行，拉伸倍数较小。调节好热拉伸处理机的温度和前后导辊的转速，确定热拉伸比，将萃取干燥后的凝胶丝在热处理机上进行超倍拉伸。

Under the action of tensile force, the loose lamella crystal with folded chain in gel fiber is gradually densified, and more and more tiemolecules with few entanglement points in fiber morphous region are gradually drawn up to

在拉伸力作用下，凝胶纤维中较为松散的折叠链片晶逐渐致密化，纤维中越来越多的具有很少缠结点的非晶区缚结分子先后被拉直而形成新的晶区，从而使

form new crystalline region, resulting in a higher degree of crystallinity and orientation of fiber and the formation of the coexistence structure of stretching chain crystallization, folding chain crystallization and tiemolecules in amorphous region. The fiber tenacity and modulus can be improved by increasing the drawing ratio. The main factors affecting the maximum draw ratio are the initial concentration of polymer, molecular weight, molecular weight distribution and drawing temperature. The gel can not be formed at too low polymer concentration (0.1%). The entanglement is numerous at too high polymer concentration, which is not conducive to stretching. The higher the molecular weight is, the higher the maximum drafting ratio is. Drawing at crystallization temperature (80 ~ 90 ℃) is beneficial to the increase of maximum drafting ratio.

纤维的结晶度、取向度得到较大提高，形成伸直链结晶、折叠链结晶和非晶区缚结分子并存的结构。提高拉伸倍数，从而提高纤维的强度和模量。影响最大拉伸倍数的主要因素有聚合物初始浓度、分子量、分子量分布和拉伸温度。聚合物浓度过低（0.1%）不能形成凝胶，聚合物浓度过高，缠结数多，不利于拉伸。分子量越高，最大拉伸倍数越大。在结晶温度（80~90 ℃）拉伸，有利于最大拉伸倍数的提高。

3 Structure/ 结构

The macromolecular orientation of fiber is shown in Figure 11-1. The orientation of UHMWPE fiber is higher than 95% and the crystallization is 85%, while those of normal PE fiber are lower and less than 60%, respectively. The UHMWPE fiber is highly crystalline and the crystallinity is typically >80%. The **crystal domain**s of UHMWPE fiber are mainly **orthorhombic crystals** with a small number of **monoclinic crystals**, highly extending along the fiber direction. The crystal domains are organized in nano- or microfibrils, which in their turn form **macrofibril**s. The larger part of the non-crystalline fraction is in the form of an **interphase** that is characterized by a high density, a high orientation and restricted mobility of the molecular chains. UHMWPE fiber has a more or less round cross-section and smooth skin.

纤维的大分子取向如图11-1所示，UHMWPE 纤维的取向高于95%，结晶达85%，而普通 PE 纤维的取向较低，结晶小于60%。UHMWPE 纤维结晶度较高，一般大于80%。UHMWPE 纤维的晶畴主要为正交晶，小部分单斜晶，沿纤维方向有很强的延伸。晶畴由纳原纤或微原纤组成，纳原纤或微原纤维形成巨原纤。非晶部分的大部分是以高密、高取向和限制分子链运动的界面相形式存在的。UHMWPE 纤维横截面接近圆形，表面光滑。

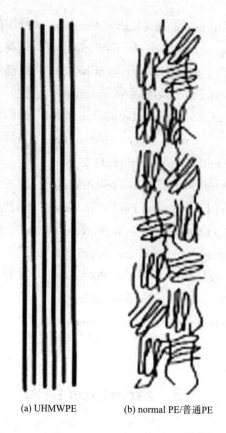

(a) UHMWPE　　(b) normal PE/普通PE

Figure 11-1　Macromolecular orientation of fiber
图11-1　纤维的大分子取向

4　Property/ 性能

Dyneema and Spectra are produced as a multifilament yarn. The **titre** of the monofilament varies about 0.3 D (0.44 dtex) to 10 D (11dtex). Tenacity of one filament may well over 5 N/tex, and the modulus can be over 120 N/tex. Staple fiber is not produced as such. Stretch broken and cut fibers are used by specialized companies. The product **portfolio** for Dyneema and Spectra at the end of 2000 is shown in Table 11-1.

以复丝纱生产Dyneema和Spectra，单丝纤度为0.44~11tex（0.3~10旦），单纤强度可以达到5 N/tex，模量可以超过120 N/tex。一般不生产短纤维，专业化公司使用拉伸断裂和切断纤维。2000年年底，Dyneema和Spectra的产品系列明细见表11-1所示。

Table 11-1 Commercially available UHMWPE filament yarns
表 11-1 商用 UHMWPE 长丝纱

Name 名称		Density 密度（kg/m³）	Denier 旦尼尔（旦）	Tenacity 强度（N/tex）	Modulus 模量（N/tex）	Elongation ratio 伸长率（%）
	Dyneema SK60	970	1	2.8	81	3.5
	Dyneema SK65	970	1	3.1	97	3.6
	Dyneema SK75	970	2	3.5	110	3.8
	Dyneema SK76	970	2	3.7	120	3.8
Toyobo	Dyneema SK60	970	1	2.8	91	3.5
	Dyneema SK71	970	1	3.5	122	3.7
Honeywell	Spectra 900	970	10	2.6	75	3.6
	Spectra 1000	970	5	3.2	110	3.3
	Spectra 2000	970	3.5	3.4	120	2.9

4.1 Density/ 密度

UHMWPE fiber has a density (970~980 kg/m³) lightly less than one, so the fiber floats on water. It is typical for highly crystalline linear PE.

UHMWPE 纤维的密度为 970~980 kg/m³，略小于1，因此，纤维浮在水面，这是高结晶线性 PE 的典型特征。

4.2 Tensile Property/ 拉伸性能

UHMWPE fiber has the highest specific tenacity and modulus among the high performance fibers at present due to the complete extension of molecular chains, high orientation and crystallization of the fiber. Its tenacity is 10~15 times that of good quality steel and 2 times that of high tenacity carbon fiber, and its modulus is second only to that of special carbon fiber grades and high modulus PBO. Elongation ratio at break is relatively low, like other high performance fibers. However, the fracture work is higher owing to the high tenacity. The performace comparison of several high-performance fibers is shown in Table 11-2. Theoretical and commercial values for tenacity and modulus of fibers are shown in Table 11-3. In contrast to the high

由于分子链完全伸展、纤维高度取向和结晶，UHMWPE 纤维是目前高性能纤维中比强度和比模量最高的纤维。其强度是优质钢的 10~15 倍，是高强碳纤维的 2 倍，模量仅次于特种碳纤维产品和高模量 PBO。与其他高性能纤维一样，纤维断裂伸长率相对较低，但由于强度高，断裂功较高。几种高性能纤维的性能对比见表 11-2。纤维的理论和实际强度和模量见表 11-3。与高拉伸强度相反，UHMWPE 纤维

tensile tenacity, UHMWPE has low compressive yield tenacity, approximately 0.1 N/tex.

具有低压缩屈服强度，约为 0.1 N/tex。

Table 11-2 Performance comparison of several high performance fibers
表 11-2 几种高性能纤维的性能对比

Property 性能	Dyneema	Kevlar		Carbon fiber 碳纤维		Glass fiber 玻璃纤维	PA66 锦纶66
	SK66	29	49	HM	LM	S-2	HT
Density 密度（g/cm³）	0.97	1.44	1.45	1.85	1.78	2.55	1.14
Tensile tenacity 拉伸强度（GPa）	3.10	2.67	2.76	2.30	3.40	2.00	0.90
Elastic modulus 弹性模量（GPa）	100	88	124	390	240	73	6
Elongation at break 断裂伸长（%）	3.8	3.6	2.5	1.5	1.4	2.0	20.0
Specific tenacity 比强度（GPa）	3.30	1.85	1.90	1.24	1.91	0.78	0.80
Specific modulus 比模量（GPa）	103	61	86	210	135	1	5

Table 11-3 Theoretical and commercial values for tenacity and modulus of fibers
表 11-3 纤维的理论和实际强度和模量

Fiber 纤维	Theoretical tenacity 理论强度		Commercial tenacity 实际强度		Theoretical modulus 理论模量		Commercial modulus 实际模量	
	GPa	N/tex	GPa	N/tex	GPa	N/tex	GPa	N/tex
UHMWPE	32	33	3.6	3.7	240	247	116	120
Aramid 芳纶	30	21	3.3	2.3	183	127	120	83
PA-6 锦纶6	32	23	0.9	0.8	142	125	6	5
PES 涤纶	28	20	1.1	0.8	125	90	14	10
PP 丙纶	18	20	0.6	0.6	34	38	6	6

The fiber tenacity can also be expressed as breaking length. The free breaking length (km) is the theoretical length of fiber, yarn or rope at which it breaks under its own weight when hanging freely. This breaking length is material

纤维的强度也可以用断裂长度来表示，自由断裂长度（km）是纤维、纱线或绳索自由悬挂时在自身重量下断裂的理论长度。

related and corresponds to the tenacity. It is independent of the thickness of the fiber or the rope. The free breaking length of Dyneema and Spectra fibers would in theory reach to a satellite's orbit.

断裂长度与材料有关，与强度相对应，与纤维或绳索细度无关。理论上，Dyneema 和 Spectra 纤维的自由断裂长度可以达到直至卫星轨道的长度。

4.3 Mechanical Property in Transverse Direction/ 横向力学性能

The mechanical properties are highly anisotropic because all the chains in the fiber are arranged in the fiber direction. The modulus and tenacity in the transverse direction are much lower than those in the fiber direction. The estimated values are shown in Table 11-4.

由于纤维中所有链都沿纤维方向排列，其力学性能是高度各向异性的。横向模量和强度远低于纤维方向，其估计值见表 11-4。

Table 11-4 Transverse properties of UHMWPE fiber
表 11-4 UHMWPE 纤维横向性能

Property 性能	Estimated value 估计值
Transverse elastic modulus 横向弹性模量（GPa）	3
Transverse compressive yield stress 横向压缩屈服应力（GPa）	0.05
Transverse tensile tenacity 横向拉伸强度（GPa）	0.03

4.4 Viscoelasticity/ 黏弹性能

PE is a viscoelastic material. The viscoelasticity of UHMWPE fiber depends significantly on temperature and loading. The tensile strength, modulus and elongation at break of UHMWPE fiber depend on temperature and strain rate. At high strain rate or at low temperature, modulus and tenacity obviously increase, which is important in ballistic protection. The UHMWPE fiber is easy to creep. The deformation increases with the increase of loading time, leading to lower modulus and higher strain at break.

PE 是一种黏弹性材料，UHMWPE 纤维黏弹性能在很大程度上取决于温度和负载。UHMWPE 纤维的拉伸强度、拉伸模量和断裂伸长率等力学性能取决于温度和应变率。在高应变率下或者在低温下，模量和强度都明显增加，这对弹道防护很重要。UHMWPE 纤维容易蠕变，

Therefore, creep is very important when the fiber is under relatively high load for a long time.

变形随负载时间的增加而增大，从而使得断裂时模量较低和应变较高。因此，当纤维长时间处于相对较高的负荷下时，蠕变就很重要。

4.5 Impact Resistance and Bulletproof/耐冲击性能和防弹性能

UHMWPE fiber has excellent impact resistance and ballistic resistance due to low density, high tenacity and modulus. It can absorb extremely high amounts of energy. This property is utilized in products for ballistic protection, cut-resistance gloves and motor helmet. It can also be used to improve the impact tenacity of carbon or glass fiber-based composites. In these applications, the high tenacity and the high energy absorption are applied. According to the energy absorption degree of the material to bullet or fragment the material bulletproof is measured. The density, tenacity, modulus and elongation at break of fiber have important influences on the bulletproof effect.

UHMWPE 纤维密度低、强度和模量高，因此，具有优异的耐冲击性和防弹性能。可吸收极高的能量，该特性应用于防弹、抗切割手套和运动头盔等产品，还可用于提高碳纤维或玻璃纤维基复合材料的冲击强度。在这些应用中，利用了高强度和高能量吸收。根据材料对弹丸或破片的能量吸收程度来衡量材料防弹性能，纤维的密度、强度、模量和断裂伸长率等对防弹效果有重要的影响。

4.6 Fatigue and Abrasion Resistance/疲劳和耐磨性能

UHMWPE fiber is the first high performance fiber with high tenacity and good tensile and bending fatigue properties comparable with the commonly used polyamide and polyester grades in ropes. Carbon fiber and glass fiber have a high modulus and a brittle breaking mode. UHMWPE fiber has a high modulus but still is flexible and has a long flex life. The good flexural fatigue resistance is related to lower compressive yield stress.

UHMWPE 纤维是第一种既有很高强度又有很好拉伸和弯曲疲劳性能的高性能纤维，其拉伸和弯曲疲劳性能与绳索中常用的聚酰胺和聚酯产品相当。碳纤维和玻璃纤维具有高模量和脆性断裂模式，UHMWPE 纤维具有高模量，但仍具有弹性和较长的弯曲寿命。良好的抗弯疲劳性能与较低的压缩屈服应力有关。

In tension fatigue testing, UHMWPE rope is repeatedly loaded in tension. It is loaded to 50% of its breaking load followed by relaxation to about 5%. In bending fatigue or flex life testing, a loaded rope is moving over two or three **sheave**s. In the test with three sheaves, two sheaves bend the rope clockwise, and the middle one rotates counter-clockwise. UHMWPE fiber is able to withstand repeated axial and longitudinal loads.

Wear resistance determines **wear and tear** and service life and is very important in ropes and gloves. UHWMPE is a well-known engineering plastic, which is specially used for its excellent **wear and abrasion resistance**. Therefore, UHWMPE fiber also has good wear resistance.

4.7 Hydrophobicity/ 拒水性能

PE is not hygroscopic and does not absorb water. The water absorption in UHMEPE fiber is negligible owing to a very low porosity. It is hydrophobic. However, UHMEPE multifilament yarns used as strands in ropes or in fabrics typically have 40% voids. Therefore, water can be absorbed among fibers. Waterproof agent is generally used to achieve water repellency. UHWMPE fiber does not swell, hydrolyze or otherwise degrade in water, sea water or moisture.

4.8 Chemical Resistance/ 耐化学性能

UHMWPE fiber is made from PE without containing any aromatic rings, amides, hydroxylic or other chemical groups that are susceptible to be attacked. UHMWPE

在拉伸疲劳试验中，UHMWPE绳反复受拉伸力作用，加载到断裂负荷的50%，然后松弛到大约5%。在弯曲疲劳或弯曲寿命试验中，一根受力绳索在两个或三个滑轮上移动，在三滑轮试验中，两个滑轮顺时针弯曲绳索、中间滑轮逆时针转动。UHMWPE纤维都能够很好地耐反复的轴向和径向载荷。

耐磨性决定磨损和使用寿命，对绳索和手套而言非常重要。UHWMPE是一种耳熟能详的工程塑料，因其优异的耐磨性而被特别使用。因此，UHWMPE纤维也具有良好的耐磨性。

PE不吸湿、不吸水。UHMEPE纤维的孔隙率很低，因此，纤维中吸水可以忽略，即是拒水的。然而，在绳索或织物中用作股线的UHMEPE复丝纱，通常有40%的孔隙，因此，纤维之间可以吸收水分。一般使用防水剂实现拒水性。UHMEPE纤维不会在水、海水或湿汽中膨胀、水解或以其他方式降解。

UHMWPE纤维由PE制成，不含任何芳香环、酰胺、羟基或其他易受攻击的化学基团。

fiber is a non-polar material without polar groups in the molecular chain. A weak interfacial layer (10~100 nm) may be produced on the surface of UHMWPE fiber under tension, showing chemical inertia. The effects of chemical reagents on UHMWPE fiber and aramid are presented in Table 11-5. It can be seen that UHMWPE fiber has strong corrosion resistance to acids, alkalis, organic solvents and most chemical reagents. UHMWPE fiber is more resistant to acid and alkali than aramid.

UHMWPE 纤维是一种非极性材料，分子链中不含极性基团，表面会在张力作用下产生一层弱界面层（10~100 nm），表现出化学惰性。化学试剂对 UHMWPE 纤维和芳纶的影响见表 11-5，可以看出，UHMWPE 纤维对酸、碱、有机溶剂和大部分化学试剂有很强的耐腐蚀能力。UHMWPE 纤维的耐酸碱性能优于芳纶。

Table 11-5　Resistance of fibers to various chemicals: 6 months immersed at ambient temperature
表 11-5　纤维对各种化学品的耐受性：在环境温度下浸泡 6 个月

Chemicals 化学试剂	UHMWPE Fiber	Aramid
Distilled water 蒸馏水	***	***
Sea water 海水	***	***
10% detergent 10% 清洁剂	***	***
Hydrochloric acid (pH=0) 盐酸	***	*
Nitric acid (pH=1) 硝酸	***	*
Glacial acetic acid 冰醋酸	***	***
Ammonium hydroxide 氢氧化铵	***	**
Sodium hydroxide (pH>14) 氢氧化钠	**	*
Petrol 汽油	***	***
Kerosene 煤油	***	***
Toluene 甲苯	***	**
Trichloromethane 三氯甲烷	***	***

Note1/注　***Unaffected/不影响, **slightly affected/轻微影响, *seriously affected/严重影响。

UHMWPE fiber has **polyolefinic** properties and is sensitive to oxidizing media. In strongly oxidization medium, fiber tenacity is quickly lost. It is stable in air.

UHMWPE 纤维有聚烯烃性质，对氧化介质敏感。在强氧化介质中，纤维会很快失去强度，在空气中较稳定。

4.9 Resistance to Light and Other Radiation/ 耐光和其他辐射性能

UHMWPE fiber has good light resistance and other radiation. Its tenacity decreases when exposed to UV light for a long time. Exposure to high-energy radiation such as electron beam or gamma radiation can lead to chain breakage and tenacity reduction.

UHMWPE 纤维具有较好的耐光和其他辐射性能，但是长时间暴露于紫外光下时，UHMWPE 纤维强度下降。暴露电子束或伽马辐射等高能辐射时，会导致链断裂和强度降低。

4.10 Electrical Property/ 电学性能

PE is an insulator without dipole characteristic groups.. UHMWPE fiber has high resistivity (volume resistivity>10^4 $\Omega \cdot m$), low dielectric constant and very low dielectric loss factor (2×10^{-4}). As-spun UHMWPE yarn contains spinning oil, which should be removed for applications requiring high electrical insulation.

PE 是一种绝缘体，没有偶极特性基团。UHMWPE 纤维具有高电阻率（体积电阻率 >10^4 $\Omega \cdot m$）、低介电常数和极低的介电损耗因数（2×10^{-4}）。所纺 UHMWPE 纱线含有纺纱油剂。因此，对于电绝缘性能要求很高的应用场合，应该去除纺纱油剂。

4.11 Acoustic Property/ 声学性能

As with all mechanical properties, the acoustic property is also anisotropic. The sound speed ($10\sim12\times10^3$ m/s) in the fiber direction is much higher than that in the transverse direction (2×10^3 m/s). The acoustic impedance is the product of density and transverse sound speed, which is close to that of water.

与所有力学性能一样，声学性能也是各向异性的。纤维方向的声速（$10\sim12\times10^3$ m/s）远高于横向（2×10^3 m/s）。声阻抗是密度和横向声速的乘积，接近于水的声阻抗。

4.12 Biological Resistance and Toxicity/ 耐生物性能和毒性

The biological resistance of UHMWPE fiber is that of high density PE. It is not sensitive to attack by microorganisms. it is regarded as biologically inert.

UHMWPE 纤维的耐生物性能与高密度 PE 一样，对微生物的攻击不敏感，是生物惰性纤维。

4.13 Thermal Resistance/ 耐热性能

The melting point of UHMWPE fiber is 144—155 ℃. The tenacity and modulus decrease at higher temperature but increase at sub-ambient temperature. There is no brittle point down to 4 K (−269 ℃), so the fiber can be used from **cryogenic** condition up to a temperature of −100 ~ −150 ℃. Short exposure to higher temperature below the melting temperature will not cause any serious loss of properties. The mechanical property is influenced by the temperature and the main reason is thought to be **chain slippag**. The breaking below 100 ℃ is within the brittle failure range.

UHMWPE 纤维熔点为 144~155 ℃，在高温下强度和模量降低，但在亚环境温度下增加。当温度降至 −269 ℃（4 K）时，纤维无脆化点，可从低温状态到 −100~−150 ℃ 使用。短暂暴露于低于熔化温度的较高温度下不会造成任何严重的性能损失。力学性能受温度的影响，主要原因为链滑移。100 ℃ 以下的断裂在脆性破坏范围内。

4.14 Shrinkage/ 收缩性能

UHMWPE fiber has high molecular extension. It does not shrink below 100 ℃, and returns to the **thermodynamiclly coiled conformation** at 120~140 ℃, resulting in shrinkage. If the fiber is restrained, the shrinkage is more significant. Table 11-6 gives an overview of the chemical and physical properties of UHMWPE fiber.

UHMWPE 纤维具有高的分子伸展特性，在 100 ℃ 以下不收缩，120~140 ℃ 回到热力学上的卷曲构象，发生收缩。如果纤维受到约束，收缩更显著。表 11-6 概述了 UHMWPE 的化学和物理性能。

Table 11-6　Overview of chemical and physical properties of UHMWPE fiber

表 11-6　UHMWPE 纤维化学物理性能概况

	Property 性能	Value 值
water and chemicals 水和化学品	Moisture regain 回潮率（%）	Zero/0
	Resistance to water 耐水性	Excellent 极好
	Resistance to acid 耐酸性	Excellent 极好
	Resistance to alkali 耐碱性	Excellent 极好
	Resistance to most chemicals 耐大部分化学品	Excellent 极好
	Resistance to UV 防紫外	Very good 很好
Thermal 热学	Melting point 熔点	144~155 ℃
	Boiling water shrinkage ratio 沸水收缩率（%）	<1
	Thermal conductivity (along fiber axis) 热传导（沿纤维轴向）[W/（m·K）]	20
	Thermal expansion coefficient 热膨胀系数（/K）	-12×10^{-6}
Electrical 电学	Resistivity 电阻（Ω·m）	$>10^4$
	Dielectric strength 介电长度（kV/cm）	900
	Dielectric constant (22 ℃, 10 GHz) 介电常数	2.25
	Loss tangent 损耗角正切	2×10^{-4}
Mechanical 力学	Axial tensile tenacity 轴向拉伸强度（GPa）	3
	Axial tensile modulus 轴向拉伸模量（GPa）	100
	Creep (22 ℃, 20% load) 蠕变（%/天）	10^{-2}
	Axial compressive tenacity 轴向压缩强度（GPa）	0.1
	Axial compressive tenacity 轴向压缩模量（GPa）	100

续表

	Property 性能	Value 值
Mechanical 力学	Transverse tensile modulus 横向拉伸强度（GPa）	0.03
	Transverse modulus 横向模量（GPa）	3

5 Application/ 应用

As UHMWPE fiber has many excellent properties and good textile processability. It can be processed by composite spinning, weaving, knitting, braiding and non-weaving, etc. UHMWPE fiber as reinforcing material can greatly reduce composite weight and improve impact tenacity and protective property of composite. Surface treatment of UHMWPE fiber and its fabric can improve its adhesion with polymer resin matrix, thereby improving the mechanical property of composite. Therefore, it has shown great advantages in the high performance fiber market, including mooring rope of offshore oil field and high performance light composite. It plays an important role in the fields of national defense, aerospace, marine defense equipment, military equipment, shipping communication, industry, civil and so on.

由于UHMWPE纤维具有众多的优异特性和纺织加工性能，可进行复合纺丝、机织、针织、编织和非织造等加工。UHMWPE纤维作为增强材料可大幅度减轻复合材料的质量、提高冲击强度和防护性能，UHMWPE纤维及其织物经表面处理可改善其与聚合物树脂基体的黏结性，从而改善复合材料的力学性能。因此，它在高性能纤维市场上，包括从海上油田的系泊绳到高性能轻质复合材料方面均显示出很大优势，在国防、航空航天、海域防备、武器装备、航运交通、工业和民用等领域发挥着举足轻重的作用。

5.1 National Defense/ 国防

Because UHMWPE fiber has good impact resistance and greater specific energy absorption, it can be made into protective fabric, helmet and bullet-proof material used in military, such as armored protective panels of helicopter, tank and ship, radar protective cover, missile cover, bullet-proof clothing, stab-proof clothing, shield and so on. Among

由于UHMWPE纤维的耐冲击性能好，比能量吸收大，在军事上可以制成防护面料、头盔、防弹材料，如直升飞机、坦克和舰船的装甲防护板、雷达防护外壳罩、导弹罩、防弹衣、防刺

them, the application of bullet-proof clothing is the most noticeable. It has the advantages of lightness and softness. UHMWPE fiber has better bullet-proof effect than aramid, and has become the main fiber occupying the American bullet-proof vest market. In addition, the specific bullet's impact load of UHMWPE fiber composite is 10 times that of steel and 2 times that of glass fiber and aramid. The bullet-proof and riot-proof helmet made of UHMWPE fiber reinforced resin composite has become a substitute for the helmet made of steel and aramid reinforced composite abroad.

衣、盾牌等，其中以防弹衣的应用最引人注目。它具有轻柔的优点，防弹效果优于芳纶，现已成为占领美国防弹背心市场的主要纤维。另外 UHMWPE 纤维复合材料的比弹击载荷值是钢的 10 倍，是玻璃纤维和芳纶的 2 倍。国外用 UHMWPE 纤维增强的树脂复合材料制成的防弹、防暴头盔已成为钢盔和芳纶增强的复合材料头盔的替代品。

5.2 Aerospace/ 航空航天

In aerospace engineering, because of its light weight, high tenacity and good impact resistance, UHMWPE fiber composite is suitable for wingtip structure of various aircraft, spacecraft structure and buoy aircraft, etc. UHMWPE fiber can also be used as the deceleration parachute for space shuttle landing and the rope for hanging heavy object on aircraft, replacing traditional steel cable and synthetic fiber cable, and its development speed is very rapid.

在航空航天工程中，由于 UHMWPE 纤维复合材料轻质高强和抗冲击性能好，适用于各种飞机的翼尖结构、飞船结构和浮标飞机等。UHMWPE 也可以用作航天飞机着陆的减速降落伞和飞机上悬吊重物的绳索，取代了传统的钢缆绳和合成纤维绳索，其发展速度异常迅速。

5.3 Industrial and Civil/ 工业用和民用

（1）Rope and Cable

The rope, cable, sail and fishing gear made of UHMWPE fiber are suitable for marine engineering and are its original use. It is widely used for loaded rope, heavy loaded rope, salvage rope, dragline, sailboat rope and fishing line, etc. The breaking length of UHMWPE rope under its own weight is 8 times that of steel rope and 2 times that of aramid. UHMWPE rope is used for fixed anchor ropes of supertanker, offshore operation platform, lighthouse, etc. It solves the problems of tenacity reduction, breakage and

（1）绳索和缆绳

UHMWPE 纤维制成的绳索、缆绳、船帆和渔具适用于海洋工程，是该纤维的最初用途，广泛用于负力绳索、重载绳索、救捞绳、拖拽绳、帆船索和钓鱼线等。UHMWPE 纤维绳索在自重下的断裂长度是钢绳的 8 倍，芳纶的 2 倍。UHMWPE 绳索用于超级油轮、海洋操作平台、灯塔

frequent replacement of the previously used ropes, which results from the corrosion of steel rope and corrosion, hydrolysis and ultraviolet degradation of nylon and polyester ropes.

（2）Sports equipment and supply

The sports goods have been made such as safety hat, ski, sail board, fishing pole, racket and parts of bicycle, glider and ultra-light aircraft etc, and their performances are superior to those made of traditional materials.

（3）Biomedical Material

UHMWPE fiber has better biocompatibility, durability and very high stability, and it does not cause **allergy**. It has been used in **clinical practice**. UHMWPE fiber is used in medical gloves and other medical measures, etc. UHMWPE fiber reinforced composite is used in **dental bracket** material, **medical implant** and **plastic suture**, etc.

（4）Industrial Material

UHMWPE fiber and its composite can be used as pressure container, conveyor belt, filter material, automobile buffer plate, etc. It can be used as wall and partition structure in building. It can be used as reinforced cement composite to improve the toughness and impact resistance of cement.

UHMWPE fiber has good dielectric property, and can be used in **antenna fairing** of **radio transmitter**, reinforcement core of optic fiber cable and so on.

等的固定锚绳，解决了以往使用钢缆遇到的锈蚀和使用锦纶、聚酯缆绳遇到的腐蚀、水解、紫外降解等引起缆绳强度降低、断裂和经常更换的问题。

（2）体育器材用品

在体育用品上已经制成安全帽、滑雪板、帆轮板、钓竿、球拍及自行车、滑翔板、超轻量飞机零部件等，其性能优于传统材料制品。

（3）生物医用材料

UHMWPE 纤维的生物相容性和耐久性都较好，并具有很高的稳定性，不会引起过敏，已在临床实践中应用。UHMWPE 纤维用于医用手套和其他医疗措施等方面，UHMWPE 纤维增强复合材料用于牙托材料、医用植入体和整形缝合线等方面。

（4）工业材料

UHMWPE 纤维及其复合材料可用作耐压容器、传送带、过滤材料、汽车缓冲板等。在建筑方面可以用作墙体、隔板结构等，用它作增强水泥复合材料可以改善水泥的韧度，提高其抗冲击性能。

UHMWPE 纤维具有良好的介电性能，可用于无线电发射装置的天线整流罩、光纤电缆加强芯等方面。

Chapter Eleven Ultra-high Molecular Weight Polyethylene Fiber/ 超高分子量聚乙烯纤维

Exercises/ 练习

1. What's the UHWMPE fiber? What properties does the fiber have?

2. Translate the following Chinese into English.

（1）UHMWPE 纤维是一种非极性材料，分子链中不含极性基团，表面会在张力作用下产生一层弱界面层（10~100 nm），纤维表面表现出化学惰性，对酸碱、有机溶剂和一般化学试剂有很强的耐腐蚀能力。

（2）UHMWPE 纤维密度低（0.96 g/cm^3）、比强度（2.7~3.8 cN/dtex）和比模量高，同时又具有优良的吸能和阻尼性能，可以用作防弹、防切割和耐冲击材料。

（3）UHMWPE 纤维及其织物的表面处理可改善其与聚合物树脂基的黏结性，从而改善复合材料的力学性能。UHMWPE 纤维作为增强材料可大幅度减轻复合材料的质量、提高冲击强度和防护性能。

（4）UHMWPE 纤维在军事、航空航天等领域发挥着重要的作用，此外，在汽车、船舶制造、医疗器械、体育运动器材等领域也有广阔的应用前景。该纤维一经问世就引起世界各国的重视，发展较快。

References/ 参考文献

[1] HEARLS J W S. High-performance fibers [M]. Cambridge:Woodhead Publishing Limited, 2001.

[2] 晏雄. 产业用纺织品 [M]. 上海：东华大学出版社，2010.

[3] 张以河. 复合材料学 [M]. 北京：化学工业出版社，2011.

[4] 姚穆. 纺织材料学 [M]. 北京：中国纺织出版社，2009.

[5] 赵刚，赵莉，谢雄军. 超高分子量聚乙烯纤维的技术与市场发展 [J]. 纤维复合材料，2011，1: 50-56.

[6] 顾超英，赵永霞. 国内外超高分子量聚乙烯纤维的生产与应用 [J]. 纺织导报，2010, 4: 52-55.

Words and Terminologies
生词和专业词汇

Chapter One   Kapok

cellulose ['seljuləus] n. 纤维素
kapok ['kepɔk] n. 木棉；木丝棉
density ['densiti] n. 密度
luster ['lʌstə] n. 光泽；v. 有光泽；发亮；使有光泽
absorption [əb'zɔːpʃn] n.（液体、气体等的）吸收
sound absorption [saund əb'zɔːpʃn] n. 吸声（性）；吸音（性）
retention [ri'tenʃn] n.（液体、热量等的）保持；维持；保留
warmth retention [wɔːmθ ri'tenʃn] n. 保暖（性）
bombacaceae ['bɔmbə'kæsiː] n. 木棉科（高大热带树）
seed fiber [siːd faibə] n. 种子纤维
epidermal cell [,epi'dəːməl sel] n. 表皮细胞
fruit fiber [fruːt faibə] n. 果实纤维
lining cell ['lainiŋ sel] n. 衬细胞
capsule shell ['kæpsjuːl ʃel] n. 胶壳；胶囊壳
cotton ginning ['kɔtn 'dʒiniŋ] n. 轧花；轧棉
crate [kreit] n. 板条箱；篓；vt. 将某物装入大木箱或板条箱中
sieve [siv] n. 滤器；筛子；漏勺；vt. 筛；过筛；滤
species ['spiːʃiːz] n. 种，物种
lignin ['lignin] n. 木质素
hemicellulose [hemi'seljuləus] n. 半纤维素
cellulosic [,selju'ləusik] adj. 纤维素的；有纤维质的
polymerization [,pɔliməraiˈzeiʃn] n. 聚合
degree of polymerization n. 聚合度
supramolecular [,sjuːprəmələ'kjulə] n. 超分子
microfibril [,maikrəu'faibril] n. 微原纤；微纤维
orientation [,ɔːriən'teiʃn] n. 取向；定向
microfibrillar orientation n. 微纤取向
perpendicular [,pəːpən'dikjələ(r)] adj. 垂直的；成直角的
parallel ['pærəlel] adj. 平行的
crystallization [,kristəlai'zeiʃn] n. 结晶（作用，过程）；结晶体；晶化
hemp [hemp] n. 大麻
morphological [,mɔːfə'lɔdʒikəl] adj. 形态（学）的

flat ribbon [flæt 'ribən] n. 扁平带状
longitudinal[ˌlɔŋgi'tjuːdinl] adj. 纵的；纵向的
natural convolution['nætʃrəl ˌkɔnvə'luːʃn] n. 天然转曲
kidney-shaped ['kidni ʃeipt] adj. 腰圆形的；肾形的
cross section['krɔːs sekʃn] n. 横截面；剖面（图）；断面（图）
lumen ['luːmin] n. 中腔
cell sap[sel sæp] n. 透明质；细胞液；核液
morphology[mɔː'fɔlədʒi] n. 形态；形态学
coarse[kɔːs] adj. 粗糙的；粗的
fine[fain] adj. 高质量的；美好的；健康的；身体很好的；细的
elliptical[i'liptikl] adj. 椭圆的；椭圆形的
thin[θin] adj. 薄的；细的；稀少的；稀疏的
thickness['θiknəs] n. 厚；厚度；粗；层
hollow degree ['hɔləu di'griː] n. 中空度
distinct[di'stiŋkt] adj. 清晰的；清楚的；明白的；明显的
linear density['liniə(r) 'densəti] n. 线密度
breaking tenacity['breikiŋ tə'næsəti] n. 断裂强度
elongation ratio at break n. 断裂伸长率
modulus ['mɔdjuləs] n. 模量
torsional [tɔːʃənəl] adj. 扭转的；扭力的
rigidity[ri'dʒidəti] n. 刚度
compression[kəm'preʃən] n. 压缩
moisture regain['mɔistʃə(r) ri'gein] n. 回潮率；回潮
dye[dai] v. 给…染色；染；n. 染料；染液
refractive index [ri'fræktiv'indeks] n. 折射率
fineness['fainnəs] n. 细度
staple fiber['steipl faibə] n. 短纤维；短纤；人造短纤维；短纤棉
twist[twist] v. 加捻；n. 捻
plastic deformation['plæstik ˌdiːfɔː'meiʃn] n. 塑性变形；永久形变；
moisture absorption ['mɔistʃə(r) əb'zɔːpʃn] n. 吸湿（性）
mercerize['məːsəraiz] v. 将（棉布或棉纱）作丝光处理
hydrophilic [ˌhaidrə'filik] adj. 亲水（性）的
crystallinity [ˌkristə'linəti] n. 结晶度；结晶性
amorphous [ə'mɔːfəs] adj. 无定形的；不规则的；无组织的
specific surface area[spə'sifik 'səːfis 'eəriə] n. 比表面积；比表面；表面积
filament ['filəmənt] n. 细丝；丝状物
spin [spin] v. 纺（纱）；纺（丝）
static['stætik] adj. 静止的；静态的
thermal conductivity['θəːml ˌkɔndʌk'tivəti] n. 导热性（率）；导热系数
dyeability [ˌdaiə'biləti] n. 染色性；可染性
affinity[ə'finəti] n. 喜好；喜爱；密切的关系；亲和性

hydrophobicity [ˌhaidrəˈfɔbiciti] n. 疏水（性）
oleophilicity [ˌəuliəuˈfiliciti] n. 亲油（性）
hydrophobic [ˌhaidrəˈfɔbik] adj. 疏水（性）的
oleophilic [ˌəuliəuˈfilik] adj. 亲油（性）的
insulation[ˌinsjuˈleiʃn] n. 隔热；隔音；绝缘；隔热（或隔音、绝缘）材料
sound insulation [saund ˌinsjuˈleiʃn] n. 隔音（性）；隔声（性）；隔音材料
spinnability [ˌspinəˈbiləti] n. 可纺性；纺丝性
cohesive force [kəuˈhiːsiv fɔːs] n. 抱合力；凝聚力；黏结力；黏合力
thick[θik] adj. 厚的；粗的；浓密的；稠密的；茂密的
evenness [ˈiːvənnəs] n. 均匀性；均匀度
hairiness [ˈheərinis] n. 毛羽
blended yarn[ˈblendid jaːn] n. 混纺纱
moisture permeability[ˈmɔistʃə(r) ˌpəːmjəˈbiliti] n. 透湿（性）
air permeability[eə(r) ˌpəːmjəˈbiliti] n. 透气（性）
moisture conductivity[ˈmɔistʃə(r) ˌkɔndʌkˈtivəti] n. 导湿（性）
porous[ˈpɔːrəs] adj. 多孔的；透水的；透气的
acid[ˈæsid] n. 酸；adj. 酸的；酸性的
acid resistance[ˈæsid riˈzistəns] n. 耐酸（性）；抗酸（性）
alkali [ˈælkəˈlai] n. 碱
alkali resistance[ˈælkəlai riˈzistəns] n. 耐碱（性）；抗碱（性）
solubility[ˌsɔljuˈbiləti] n. 溶（解）度；（可）溶（性）
acidic[əˈsidik] adj. 酸性的；含酸的；很酸的
alkaline[ˈælkəlain] adj. 碱性的；含碱的
solvent[ˈsɔlvənt] n. 溶剂；adj. 有溶解力的；可溶解的
dissolve[diˈzɔlv] v. 溶；使（固体）溶解
hydrochloric acid[ˌhaidrəˌklɔrik ˈæsid] n. 盐酸
nitric acid[ˌnaitrik ˈæsid] n. 硝酸
acetic acid[əˌsiːtik ˈæsid] n. 乙酸；醋酸
formic acid[ˌfɔːmik ˈæsid] n. 甲酸；蚁酸
sodium hydroxide[ˈsəudiəm haiˈdrɔksaid] n. 氢氧化钠；苛性钠；烧碱
sulphuric acid[sʌlˌfjuərik ˈæsid] n. 硫酸
dilute[daiˈluːt] v. 稀释；冲淡；削弱；降低；adj. 稀释了的；稀的
concentrated[ˈkɔnsntreitid] adj. 浓（缩）的
concentration[ˌkɔnsnˈtreiʃn] n. 浓度；集中；聚集；专注；重视
dissolution[ˌdisəˈluːʃn] n. 溶解
insoluble[inˈsɔljəbl] adj. 无法解决的；不能解释的；不能溶解的；不溶的
soluble[ˈsɔljəbl] adj. 可溶的；可解决的；可解的
thermal stability[ˈθəːml stəˈbiləti] n. 热稳定（性）；抗热（性）
degradation[ˌdegrəˈdeiʃn] n. 降解
degrade[diˈgreid] v. （使）降解；分解；降低
thermal resistance[ˈθəːml riˈzistəns] n. 耐热（性）；热阻；热阻抗
antibacterial[ˌæntibækˈtiəriəl] adj. 抗菌的；n. 抗菌剂
mould[məuld] n. 霉；霉菌

broad spectrum[ˌbrɔːd spektrəm] adj. 广谱的；效用广泛的
antimicrobial [ˌæntimaiˈkrəubiəl] n. 抗菌剂；杀菌剂；adj. 抗菌的
reproduction[ˌriːprəˈdʌkʃn] n. 生殖；繁殖；复制；再版
anaerobic bacteria[ˌænəˈrəubik bækˈtiəriə] n. 厌氧细菌；厌气细菌
mite[mait] n. 螨虫
flavonoid[ˈfleivənɔid] n. 类黄酮
triterpene [traiˈtəːpiːn] n. 三萜；三萜类；三萜类化合物
gram-positive[græm ˈpɔzətiv] n. 革兰氏阳性；革兰氏阳性菌；阳性
escherichia coli [ˌeʃəˈrikjə kəulai] n. 大肠杆菌
gram-negative[græm ˈnegətiv] n. 革兰氏阴性；革兰氏阴性菌；阴性
staphylococcus aures [ˌstæfiləˈkɔkəs ˈɔːriə] n. 金黄色葡萄球菌
softness[ˈsɔftnəs] n. 柔软；温柔
composite[ˈkɔmpəzit] adj. 合成的；混成的；复合的；n. 复合材料；混合物
filler[ˈfilə(r)] n. 填充物，填料
reinforcing material[ˌriːinˈfɔːsiŋ məˈtiəriəl] n. 增强材料
clothing[ˈkləuðiŋ] n. 衣服；（尤指某种）服装
home textile[həum ˈtekstail] n. 家纺；家纺用品；家用纺织品
apparel[əˈpærəl] n.（商店出售的）衣服；服装；（尤指正式场合穿的）衣服
viscose[ˈviskəuz] n. 黏胶
handle[ˈhændl] n. 手感
stuffing material[ˈstʌfiŋ məˈtiəriəl] n. 填充材料
bedding [ˈbediŋ] n. 铺盖；卧具；寝具
upholstery [ʌpˈhəulstəri] n. 家具装饰用品业；垫衬物
pillow [ˈpiləu] n. 枕头
mattress[ˈmætrəs] n. 床垫
cushion[ˈkuʃn] n. 软垫；坐垫；靠垫
antibacterial activity [ˌæntibækˈtiəriəl ækˈtivəti] n. 抗菌活性；抗菌性
buoyant[ˈbɔiənt] adj. 漂浮的；能够漂起的；有浮力的
life preserver [ˈlaif prizəːvə(r)] n. 救生用具
buoyancy[ˈbɔiənsi] n. 浮力；轻快；轻松的心情
lamination [ˌlæmiˈneiʃən] n. 层压；迭片结构；薄板
wadding[ˈwɔdiŋ] n.（柔软的）填料；填絮；衬垫
melt [melt] v.（使）熔化；n. 熔体
hot-melt[hɔt melt] adj. 热熔的
compression resistance[kəmˈpreʃn riˈzistəns] n. 耐压（性）
fiber assembly[faibə əˈsembli] n. 纤维束；纤维集合体
water repellent[ˈwɔːtə ripelənt] adj. 防水的；抗水性的；疏水的
heat insulation[hiːt ˌinsjuˈleiʃn] n. 隔热；绝热；热绝缘
flexural [flekʃərəl] adj. 弯曲的；曲折的
illumination[iˌluːmiˈneiʃn] n. 照明；光源；彩灯

Chapter Two Natural Colored Cotton Fiber

boll [bəul] n. 棉铃
boll opening [bəul 'əupəniŋ] n. 吐絮
pigment ['pigmənt] n. 色素；颜料 vt. 给……着色 vi. 呈现颜色
deposit[di'pɔzit] v. 使沉积；使沉淀；n. 订金；放置；使淤积
bleach[bliːtʃ] v.（使）变白；漂白；退色；n. 漂白剂
unicellular[ˌjuːni'seljələ(r)] adj. 单细胞的
variety[və'raiəti] n. 品种
yield [jiːld] n. 产量；收益 vt. 屈服；出产；产生
pectin['pektin] n. 果胶
shade [ʃeid] vi.（颜色、色彩等）渐变；n. 阴影；（照片等的）明暗度；
infrared spectra[ˌinfrə'red 'spektrə] n. 红外光谱
stretching vibration['stretʃiŋ vai'breiʃn] n. 伸缩振动；弹性振动
absorption band[əb'zɔːpʃn bænd] n. 吸收（光、谱、光谱）带
spiral['spairəl] n. 螺旋形；adj. 螺旋形的；v. 螺旋式上升（或下降）
microstructure [ˌmaikrə'strʌktʃə(r)] n. 微观结构；显微结构
concentric[kən'sentrik] adj. 同心的
cylinder['silində(r)] n. 圆柱；圆柱体
primary wall['praiməri wɔːl] n. 初生壁
secondary wall['sekəndri wɔːl] n. 次生壁
horizontal[ˌhɔri'zɔntl] adj. 水平的；横向的；n. 水平位置；水平线；横切面
vertical['vəːtikl] adj. 竖的；垂直的；直立的；纵向的；n. 垂直位置；垂直线
tabular ['tæbjulə] adj. 扁平的
upper half mean length n. 上半部平均长度
span length[spæn leŋθ] n. 跨距长度
micronaire n. 马克隆值；棉纤维马克隆尼值；纤维细度麦克隆值
uniformity[ˌjuːni'fɔːməti] n. 整齐度；均匀性；均匀度
short fiber ratio n. 短绒率
neps [neps] n. 棉结
lint ratio [lint 'reiʃiəu] n. 分衣率
resilience [ri'ziliəns] n. 弹性；弹力
wrinkle['riŋkl] n.（布或纸上的）皱褶；皱痕；v. 皱起；（使）起皱褶
launder['lɔːndə(r)] v. 洗熨（衣物）
iron['aiən]v.（用熨斗）熨；烫平；n. 熨斗
durable-press finishing n. 免烫整理
cohesion[kəu'hiːʒn] n. 黏合；结合；凝聚性
breakage['breikidʒ] n. 断头；断裂
fuzziness['fʌzinəs] n. 起毛；绒毛的特性
impurity[im'pjuərəti] n. 杂质
sterile seed['sterail siːd] n. 不育种子；不孕籽粒

chip[tʃip] n. 碎屑；碎片；切片
defect['diːfekt] n. 疵点；缺陷
hydrophilicity [ˌhaidrə'filicity] n. 亲水（性）
dimensional stability[dai'menʃənl stə'biləti] n. 尺寸稳定性
shrink[ʃriŋk] v. 收缩；（使）缩水；皱缩
deform[di'fɔːm] v. 改变…的外形；损毁…的形状
shrinkage ratio['ʃriŋkidʒ 'reiʃiəu] n. 收缩率
antistatic [ˌænti'stætik] adj. 防静电（性）的；抗静电（性）的
static electricity['stætik iˌlek'trisəti] n. 静电
ultraviolet[ˌʌltrə'vaiələt] adj. 紫外线的；利用紫外线的；n. 紫外线辐射；紫外光
transmittance [trænz'mitəns] n. 透射（比）；透过（率）；透光（率）
color fastness['kʌlə(r) 'fɑːstnəs] n. 色牢度；牢度；染色牢度
mildew ['mildjuː] n. 霉；霉病 vt. 使发霉 vi. 发霉；生霉
bacterium [bæk'tiəriəm] n. 细菌；复数 bacteria[bæk'tiəriə]
scouring['skauəriŋ] n. 煮练；精练；洗毛
adhesive [əd'hisiv] n. 黏合剂；胶黏剂；adj. 黏结性的；黏着的；有附着力的
pectinase ['pektineis] n. 果胶酶
degreasing cotton[ˌdiː'griːsiŋ 'kɔtn] n. 脱脂棉
absorbent cotton[əbˌzɔːbənt 'kɔtn] 脱脂（吸水）棉；药棉
ammonium oxalate [ə'məuniəm 'ɔksəleit] n. 草酸铵；草胺酸
extractant [ik'stræktənt] n. 萃取剂；提取剂；提炼物；分馏物
wicking height ['wikiŋ hait] n. 芯吸高度；吸水高度；毛细升高值
immerse[i'məːs]v. 浸泡；使浸没于……
fade[feid] v. （使）变淡；（使）褪色；（使）变暗
household textile['haushəuld 'tekstail] n. 家用纺织品
casual['kæʒuəl] n. 休闲服装
hybridization[ˌhaibridai'zeiʃn] n. 杂化；杂混；杂交
transmissibility [trænzˌmisə'biləti] n. 遗传性；传递性；传导率
stain[stein]v. 沾色；玷污；留下污渍；给……染色；n. 污点；污渍；染色剂
discolor[dis'kʌlə] v. 变色；褪色；使变色；使褪色
oxidant ['ɔksədənt] n. 氧化剂
reducer [ri'djuːsə] n. 还原剂
enzyme['enzaim] n. 酶
penetrant ['penitrənt] n. 渗透剂；adj. 渗透的
detergent [di'təːdʒənt] n. 清洁剂；去垢剂
soften['sɔf(ə)n] v. （使）变软；（使）软化；（使）柔和

Chapter Three Bamboo Fiber

abrasion resistance[ə'breiʒn ri'zistəns] n. 耐磨（性）；抗磨性（力）
antibiosis [ˌæntibai'əusis] n. 抗菌（性）；抗生（性）；抗生作用；抗菌作用
bacteriostasis[bækˌtiriə'stesis] n. 抑菌（性）；抑菌作用；抑（制）细菌法
anti-mite ['ænti mait] n. 除螨（性）；防螨（性）
deodorization[diˌɔdəri'zeʃən] n. 防臭（性）；除臭（性）；脱臭（作用或过程）
anti-UV ['æntijuː 'viː] n. 防紫外线（性）；抗紫外线（性）
perennial [pə'reniəl] adj. 多年生的；常年的；四季不断的；反复的 n. 多年生植物
gramineous [grə'miniəs] adj. 禾本科的；草的；似草的
renewability [riˌnjuːə'biləti] n. 可再生（性）
plywood ['plaiwud] n. 夹板，胶合板
pulp [pʌlp] n. 浆状物；纸浆；v. 将……捣成浆
officinal [ə'fisin(ə)l] n. 成药 adj. 药用的；成药的；依照药方的
cylindrical [sə'lindrikl] adj. 圆柱形的；圆柱体的
bamboo stalk[ˌbæm'buː stɔːk] n. 竹竿；竹柄；竹茎
node [nəud] n. 节点
internode ['intənəud] n. 节间
sheath [ʃiːθ] n. 鞘；护套；叶鞘
nodal diaphragm['nəudəl 'daiə'fræm] n. 节隔膜
regenerated [ri'dʒenəreitid] adj. 再生的 v. 再生；革新
scour['skauə(r)]vt. 洗涤；冲洗
degum [di'gʌm] vt. （使）脱胶；使去胶
opening ['əupən] n. 开松
carding['kaːdiŋ] n. 梳理
hydrolysis[hai'drɔlisis] n. 水解
alkalization [ˌælkəlinai'zeifən] n. 碱处理（法）；碱处理法；碱化（作用）
refine[ri'fain] v. 精练；提纯；去除杂质
solidification [ˌsɔlidifi'keiʃən] n. 凝固；固化；浓缩；凝结过程；凝固作用
alkalize ['ælkəlaiz] v. （使）碱化
sulphonate ['sʌlfəneit]vt. 磺化
filtration[fil'treiʃn] n. 过滤
ripen['raipən] v. （使）成熟
filter['filtə(r)] n. 过滤器；v. 过滤；筛选
plasticize['plæstisaiz] v. 使增塑；使塑化；使可塑
washability [ˌwɔʃə'biliti] n. 可（耐）洗性
bast [bæst] n. 韧皮部；树的内皮
flax [flæks] n. 亚麻；亚麻纤维
jute [dʒuːt] n. 黄麻；黄麻纤维
diffraction [di'frækʃən] n.（光，声等的）衍射，绕射
ramie ['ræmi] n. 苎麻；（纺织等用的）苎麻纤维

crystal['krɪstl] n. 晶体；结晶
microcrystalline [ˌmaɪkrəʊ'krɪstəlɪn] adj. 微晶的；n. 微晶
microfiber [ˌmaɪkrəʊ'faɪbə] n. 微纤维
alternately [ɔːl'tɜːnɪtli] adv. 交替地；轮流地
transverse['trænzvɜːs] adj. 横（向）的；横断的；横切的；n. 横断面
pointed['pɔɪntɪd] adj. 尖的；有尖头的；
spindle['spɪndl] n. 纺锤；轴；vi. 装锭子于；长成细长茎
micro-groove['maɪkrəʊgruːv] n. 微沟槽
micro-crack['maɪkrəʊ kræk] n. 微裂痕；微裂纹；微裂缝
serrated[sə'reɪtɪd] adj. 锯齿状的
petal['petl] n. 花瓣
processability [prə'sesə'bɪləti] n. 可加工（性）
wearability [ˌwerə'bɪləti] n. 服用性；可穿着性；穿着性能；耐磨损性
rupture work['rʌptʃə(r) wɜːk] n. 断裂功
hygroscopic [ˌhaɪgrə'skɒpɪk] adj. 吸湿的；湿度计的；易潮湿的
uniformly['juːnɪfɔːmli] adv. 均匀地；一致地；一律；无变化地
absorbency [əb'sɔːbənsi] n. 吸收（性）
dyestuff['daɪstʌf] n. 染料；染色剂
swell [swel] v. 膨胀
diffusion [dɪ'fjuːʒən] n. 扩散；传播；漫射
anti-insect ['ænti 'ɪnsekt] n. 防虫
anti-odor['ænti 'əʊdə] n. 防臭
quinone ['kwɪnəʊn] n. 醌
allergy['ælədʒi] n. 过敏反应
sterilize['sterəlaɪz] v. 灭菌；杀菌
linen['lɪnɪn] n. 亚麻；亚麻纤维；亚麻织品
strain [streɪn] n. 菌种；应力；张力
bacillus subtilis n. 枯草芽孢杆菌
candida albicans n. 白色念珠菌
aspergillus niger n. 黑曲霉菌
adsorption [æd'sɔːpʃən] n. 吸附（性）；吸附作用
sodium['səʊdiəm] n. 钠
chlorophyll['klɒrəfɪl] n. 叶绿素
adsorb[əd'zɔːb]v. 吸附（液体、气体等）
formaldehyde[fɔː'mældɪhaɪd] n. 甲醛
benzene['benziːn] n. 苯
toluene ['tɒljuiːn] n. 甲苯
ammonia[ə'məʊniə] n. 氨；氨水
anion ['ænaɪən] n. 负离子；阴离子
irritation[ˌɪrɪ'teɪʃn] n. 刺激
comfort['kʌmfət] n. 舒适（性）；舒服
degrease [diː'griːs]vt. 脱脂；去除……的油污
deacidification [diːəˌsɪdəfə'keɪʃən] n. 脱酸作用；（油）蒸馏中和

179

deproteinization [di'prəuti:nai'zeiʃn] n. 脱蛋白作用；除蛋白
biodegradability [baidi'greidə'biləti] n. 生物降解性
decompose[ˌdi:kəm'pəuz] v.（使）分解
carbon dioxide[ˌka:bən dai'ɔksaid] n. 二氧化碳
microorganism[ˌmaikrəu'ɔ:ənizəm] n. 微生物
sludge[slʌdʒ] n. 烂泥；淤泥
lustrous['lʌstrəs] adj. 柔软光亮的；有光泽的
drapability [ˌdreipə'biləti] n. 悬垂性
stiffness[stifnəs] n. 刚度；僵硬；硬度
sanitary ['sænit(ə)ri] adj. 卫生的；清洁的
feel[fi:l] n. 手感
sterilization[ˌsterəlai'zeiʃ(ə)n] n. 灭菌
penetrate['penətreit] v. 渗透；穿过；进入

Chapter Four Milk Protein Fiber

semi-synthetic['semi sin'θetik] adj. 半合成的
copolymerize [kəu'pɔləməraiz] v. 共聚合
wet spinning[wet 'spiniŋ] n. 湿纺
casein['keisi:in] n. 酪蛋白
polyacrylonitrile ['pɔli,ækriləu'naitril] n. 聚丙烯腈
polyvinyl alcohol[pɔlivainl 'ælkəhɔ:l] n. 聚乙烯醇
aesthetic[es'θetik] adj. 美的；n. 美观
fitness['fitnəs] n. 健康；健壮
graft[gra:ft] n. 接枝；接穗；嫁接
graft copolymerization n. 接枝共聚合
trademark['treidma:k] n. 商标
surgical suture['sə:dʒikl 'su:tʃə(r)] n. 手术缝合；手术缝合线；外科缝线
macromolecule ['mækrəumɔlikju:l] n. 大分子；高分子
linear['liniə(r)] adj. 线性的；直线的；线状的
fibrin['faibrin] n. 纤维蛋白
globular['glɔbjələ(r)] adj. 球状的；球形的
globulin['glɔbjuin] n. 球蛋白
soybean['sɔi,bi:n] n. 大豆；黄豆
flexibility[,fleksə'biləti] n. 柔韧性；灵活性；弹性；适应性
intermolecular[,intəmə'lekjulə] adj.（作用于）分子间的
intermolecular force [,intərmə'lekjələr fɔ:s] n. 分子间作用力分子间力
peptide linkage['peptaid 'liŋkidʒ] n. 肽键
hydrogen bond['haidrədʒən bɔnd] n. 氢键
polar group['pəulə(r) gru:p] n. 极性基团；极性基
colloidal solution[kə'lɔidəl sə'lu:ʃn] n. 胶体溶液
polypeptide [pɔli'peptaid] n. 多肽；多肽链
solidify[sə'lidifai] v.（使）凝固；变硬
strand [strænd] n. 丝条
concentrate['kɔnsntreit] v.（使）浓缩；n. 浓缩物
isolate['aisəleit] v. 使（某物质）分离；使离析
dialysis [,dai'æləsis] n. 渗析
salt out[sɔ:lt aut] v.（使）盐析，加盐分离
purify['pjuərifai]v. 提纯；精练
knead [ni:d]vt. 糅合
filtrate['filtreit] n. 滤液；v. 过滤
deaerate [,di:eə'reit] v. 脱泡
stretch[stretʃ] v. 牵伸
decompress[,di:kəm'pres] v.（使）减压

181

centrifuge['sentrifju:dʒ] n. 离心机
centrifugal[ˌsentri'fju:gl] adj. 离心的
emulsion[i'mʌlʃn] n. 乳状液；乳浊液
semipermeable membrane [semi'pə:mitəb(ə)l 'membrein] n. 半透膜；半透性膜
inorganic[ˌinɔː'gænik] adj. 无机的
magnesium sulfate[mæg'ni:ziəm 'sʌlfeit] n. 硫酸镁
sodium chloride[ˌsəudiəm 'klɔːraid] n. 氯化钠；食盐
precipitate[pri'sipiteit] v. 使沉淀；n. 沉淀物；析出物
deionized [di'aiənaizd] adj. 去离子的
deaeration [ˌdiːeə'reiʃən] n. 脱泡；通风
pilling['piliŋ] n. 起球；结绒
broken end['brəukən end] n. 断头；断头疵；断经
spinneret ['spinəret] n. 喷丝头
orifice ['ɔrifis] n. 喷丝孔
gear[giə(r)] n. 齿轮
metering pump['miːtəriŋ pʌmp] n. 计量泵
as-spun fiber[æz spʌn faibə] n. 原丝；初生纤维
wind[waind] v. 卷绕
cone[kəun] n. 筒子
nascent fiber['næsnt faibə] n. 初生纤维
align[ə'lain] v. 排列；排整齐；使成一条直线
heat-set[hiːt set] v. 热定形；热固化
denatured protein[diː'neitʃəd 'prəutiːn] n. 变性蛋白
permanent ['pəːmənənt] adj. 永久的；永恒的；长久的
crosslink[krɔs 'liŋk] n. 交联；v. 横向连接（交叉耦合；交联）
additive['ædətiv] n. 添加剂
disperse[di'spəːs] v. （使）分散
aggregation[ˌægri'geiʃn] n. 聚集，集合，聚集体；集合体；聚合作用；凝聚
uneven[ʌn'iːvn] adj. 不均匀；不规则的
evenly['iːvnli] adv. 均匀地；规则地
catalyst['kætəlist] n. 催化剂
crystalline['kristəlain] adj. 结晶的；水晶制的；晶状的
amino[ə'miːnəu] adj. 氨基（的）
amino acid[əˌmiːnəu 'æsid] n. 氨基酸
peptide bond['peptaid bɔnd] n. 肽键
dispersed[di'spəːst] adj. 分散的；散布的
side group[said gruːp] n. 侧基；侧基团
crystallize['kristəlaiz] v. （使）形成晶体；结晶
grain size[grein saiz] n. 粒径；颗粒大小
dumbbell['dʌmbel] n. 哑铃
profiled fiber['prəufaild faibə] n. 异形纤维
concave and convex[kɔn'keiv ənd 'kɔnveks] n. 凹凸
dehydration[ˌdiːhai'dreiʃən] n. 脱水

curly['kəːli] adj. 有卷发（或毛）的；卷曲状的
yellowish['jeləuiʃ] adj. 微黄色的；发黄的
humidification ['hjuːmidi'fikeifən] n. 湿润状况；增湿（作用）；湿润（作用）
moisturize['mɔistʃəraiz] v. 使皮肤湿润；（用脂膏）滋润
friction coefficient['frikʃn ˌkəui'fiʃnt] n. 摩擦系数；摩擦因数
crimp[krimp] v.（使）卷曲；n. 卷曲
mass specific resistance[mæs spə'sifik ri'zistəns] n. 质量比电阻
acrylic fiber[ə'krilik faibə] n. 聚丙烯腈纤维；腈纶；腈纶
cationic ['kætaiənik] adj. 阳（正）离子的
reactive dye[ri'æktiv dai] n. 活性染料；作用（于被染织物）染料
fluffy['flʌfi] adj. 松软的；绒毛般的；轻软状的
three-dimensional[ˌθriː dai'menʃənl] adj. 三维的；立体的
aesthetic[iːs'θetik] adj. 审美的；美学的；n. 美感；审美观；美学；外观
anti-fuzzing ['æntifʌziŋ] n. 抗起毛
anti-pilling['ænti 'piliŋ] n. 抗起球；球纤维
light resistance[lait ri'zistəns] n. 耐光性；光电阻
healthcare['helθkeə] n. 医疗保健；医疗卫生
gel[dʒel] n. 凝胶
buff[bʌf] n. 浅黄
heat resistance[hiːt ri'zistəns] n. 耐热性
yellow['jeləu] v.（使）变黄；adj. 黄色的；n. 黄色
moist[mɔist] adj. 微湿的；湿润的
chemical stability['kemikl stə'biləti] n. 化学稳定性
chloride['klɔːraid] n. 氯化物
large-scale[ˌlaːdʒ 'skeil] adj. 大规模的；大批的；大范围的
batch production[bætʃ prə'dʌkʃn] n. 成批（分批，间歇）生产
cashmere['kæʃmiə(r)] n. 羊绒
woven fabric['wəuvn 'fæbrik] n. 机织面料；机织物；机织布
knitted fabric ['nitid 'fæbrik] n. 针织面料；针织物；针织布
ribbed fabric ['ribd 'fæbrik] n. 罗纹面料；罗纹织物
underwear['ʌndəweə(r)] n. 内衣；衬衣
toughness[tʌfnəs] n. 韧性
gorgeous['gɔːdʒəs] adj. 非常漂亮的；绚丽的；华丽的
cheongsam[tʃɔŋ'sæm] n. 旗袍
fashion['fæʃn] n.（衣服、发式等的）流行款式
elastic[i'læstik] adj. 弹性的
polyurethane [ˌpɔli'juərəθein] n. 氨纶；聚氨酯纤维
blazer['bleizə(r)] n. 运动上衣
workout clothes['wəːkaut kləuðz] n. 运动服；健身服
breathable['briːðəbl] adj. 透气的
texture['tekstʃə(r)] n. 质地；手感
luxurious[lʌg'ʒuəriəs] adj. 华贵的；奢侈的
heat preservation[hiːt ˌprezə'veiʃn] n. 保温

elasticity [ˌiːlæˈstisəti] n. 弹性；弹力
allergic[əˈləːdʒik] adj. 过敏的
nonwoven fabric [ˈnəunˈwəuvn ˈfæbrik] n. 非织造布
bandage[ˈbændidʒ] n. 绷带；v. 用绷带包扎
gauze[gɔːz] n. 薄纱；纱布
pad [pæd] n.（吸收液体、保洁或保护用的）软垫；护垫；垫状物
diaper[ˈdaipə(r)] n. 尿布；尿片
deodorant[diˈəudərənt] adj. 防臭的
tickle[ˈtikl] n. 刺痒感

Chapter Five Soybean Protein Fiber

polysaccharide [,pɔli'sækəraid] n. 多糖；多聚糖
remains [ri'meinz] n. 剩余物；残留物
soybean meal ['sɔi,bi:n mi:l] n. 豆粕
globular protein ['glɔbjələ(r) 'prəuti:n] n. 球状蛋白质；球形蛋白
glycinin ['glaisinin] n. 大豆球蛋白
debug [,di:'bʌg] v. 调试
nonpolar [nəun 'pəulə(r)] adj. 非（无）极性的
imidic acid ['imidik æsid] n. 乙氨酸
lactamic acid ['læktəmik æsɪd] n. 丙氨酸
cysteine ['sistii:n] n. 半胱氨酸
cystine ['sisti:n] n. 胱氨酸
spherical ['sferikl] adj. 球形的；球状的
spatial ['speiʃl] adj. 空间的
stir [stə:(r)] v. 搅动；搅和；搅拌
coagulation [kəu,ægju'leiʃn] n. 凝固（作用）；凝结（物）；凝聚（作用）
draft [drɑ:ft] v. 牵伸
acetalization ['æsitælizeifən] n. 缩醛化；缩醛作用
curl [kə:l] v.（使）卷曲
crimp-set [krimp set] v. 卷曲定型
hydroxyl [hai'drɔksil] n. 羟基；氢氧基
carboxyl [kɑ:'bɔksil] n. 羧基
sulfur ['sʌlfə(r)] n. 硫；vt. 用硫黄处理
salt bond [sɔ:lt bɔnd] n. 盐键；离子键
disulfide bond [di'sʌlfaid bɔnd] n. 双硫键；二硫键
amide [ə'maid] n. 酰胺；氨（基）化物
ester ['estə(r)] n. 酯
acetalation ['æsitæleifən] n. 缩醛化（作用）
crosslinkage ['krɔsliŋkidʒ] n. 交联
network structure ['netwə:k 'strʌktʃə(r)] n. 网状结构
crimpness ['krimpnis] n. 卷曲度；卷曲性
loop strength [lu:p streŋθ] n. 钩接强力
knot strength [nɔt streŋθ] n. 打结强力
hook strength [huk streŋθ] n. 钩接强力
coil [kɔil] n. 线圈；vi. 卷成圈状；vt. 盘绕，把……卷成圈
unevenness [ʌn'i:vənəs] n. 不均匀；不齐；不平
touch [tʌtʃ] n. 手感；触感；触觉；v. 触摸；碰；接触
bulkiness ['bulkinis] n. 蓬松性
plumpness ['plʌmpnis] n. 丰满度；饱满度；肥胖度

crease[kriːs] n. 褶痕；皱痕；褶缝；褶线；v. 弄皱；压褶；(使)起褶子
crease resistance[kriːs ri'zistəns] n. 耐皱性；抗皱性；耐折性
anisotropic [ˌænaisə'trɔpik] adj. 各向异性的
birefractive [bai'ri'fræktiv] adj. 双(重)折射的
birefringence [bai'ri'fridʒen] n. 双(光)折射；二次光折射；折射率
rough[rʌf] adj. 粗糙的
stiff[stif] adj. 硬的；挺的；不易弯曲的
direct dye[də'rekt dai] n. 直接染料
antiseptic[ˌænti'septik] adj. 杀菌的；无菌的；防腐的；n. 防腐剂；抗菌剂
anti-inflammatory[ˌænti in'flæmətri] adj. 消炎的；抗炎的
therein[ˌðeər'in] adv. 在那里；在其中(指提及的地点、物体、文件等)
compatible[kəm'pætəbl] adj. 可共用的；兼容的；可共存的
chemical resistance['kemikl ri'zistəns] n. 耐(抗)化学性；耐化学品能力
oxidation-reduction[ˌɔksi'deiʃn ri'dʌkʃn] n. 氧化—还原(作用)
reductant [ri'dʌktənt] n. 还原剂；强力还原剂
sodium thiosulfate ['səudiəmˌθaiəu'sʌlfeit] n. 硫代硫酸钠；大苏打
hypochlorite [ˌhaipəu'klɔːrait] n. 次氯酸盐；低氧化氯
sodium hypochlorite ['səudiəmˌhaipəu'kbːrait] n. 次氯酸钠；漂白水
hydrogen peroxide[ˌhaidrəgən pə'ɔksaid] n. 过氧化氢；双氧水
wrinkle resistance['riŋkl ri'zistəns] n. 抗皱性能

Chapter Six   Modified Wool Fiber

entangle[in'tæŋgl] vt. 使纠缠；缠住；卷入；使混乱
directional frictional effect[də'rekʃənl 'frikʃən(ə)l i'fekt] n. 定向摩擦效应
scale[skeil] n. 鳞片；v. 去鳞
felting ['feltiŋ] n. 毡缩；毡呢；v. 把……制成毡；用毡覆盖
prickle ['prikl] n. 刺；刺痛；针刺般的感觉；vt. 刺；戳；使感到刺痛
washable['wɔʃəbl] adj. 可洗的；耐洗的；n. 可洗的布料（或衣服）
wearable['weərəbl] adj. 穿戴舒适的；可穿戴的；适于穿戴的
maintenance['meintənəns] n. 保养；维护；维持；保持
count[kaunt] n. 支数
premium['priːmiəm] adj. 高昂的；优质的；n. 保险费；额外费用；附加费
shrinkproof ['shriŋkpruf] adj. 不缩的；防缩的；n. 防缩（性）；防毡缩（性）
silky['silki] adj. 丝绸的；柔软光洁的；柔和的；轻柔的；丝制的
strip[strip] vt. 除去；剥去（一层）；（尤指）剥光；剥夺
resin['rezin] n. 树脂；合成树脂；vt. 用树脂处理；涂擦树脂于……
deposition[ˌdepə'ziʃn] n. 沉积（物）；沉淀（物）
plasma ['plæzmə] n. 等离子体
sodium dichloroisocyanurate n. 二氯异氰尿酸钠
potassium permanganate [pə'tæsiəm pəmæŋɡənət] n. 高锰酸钾
persulfate [pə:'sʌlferit] n. 过硫酸盐
sodium persulfate ['səudiəm pə:'sʌlferit] n. 过硫酸钠；过二硫酸钠
chlorination[ˌklɔːri'neiʃn] n. 氯化（作用）；加氯消毒
chlorine['klɔːriːn] n. 氯；氯气
even['iːvn] adj. 均匀的；均等的；平滑的；平坦的；（数量等）变化不大的
inhibitor[ɪn'hibitə(r)] n. 抑制剂；阻聚剂
chlorinate['klɔːrineit] v. 氯化；（在水中）加氯消毒
chlorinating agent['klɔːrineitiŋ 'eidʒənt] n. 氯化剂
anti-felting['ænti 'feltiŋ] n. 防毡缩（性）
hypochlorous acid n. 次氯酸
cortex['kɔːteks] n. 皮层；皮质；（尤指）大脑皮层
peroxymonosulfuric acid n. 过氧硫酸；过一硫酸
neutral['njuːtrəl] adj. 中性的
agitate['ædʒiteit] v. 搅动；摇动（液体等）
saturated['sætʃəreitid] adj. （溶液）饱和的；深的；浓的；湿透；v. 使湿透
corrosion[kə'rəuʒn] n. 腐蚀；侵蚀
manganese['mæŋɡəniːz] n. 锰
uniform['juːnifɔːm] adj. 一致的；统一的；一律的；n. 制服；校服
calcium sulfate['kælsiəm 'sʌlfeit] n. 硫酸钙
proteinase['prəutiːneis] n. 蛋白水解酶；蛋白分解酶；蛋白分解

catalytic[ˌkætə'litik] adj. 起催化作用的；有催化性的；催化性的
endonuclease [ˌendəu'njuːklieit] n. 核酸内切酶；内切核酸酶
exonuclease [ˌeksəu'njuːklieit] n. 外切核酸酶
diffuse[di'fjuːz] v. 扩散；使分散；adj. 扩散的
hydrolyze ['haidrəˌlaiz] v. 水解；使水解
catalysis[kə'tæləsis] n. 催化
contamination [kənˌtæmi'neiʃn] n. 污染
itchy ['itʃi] adj. 发痒的；渴望的
charged[tʃɑːdʒd] adj. 带电的
positive ion['pɔzətiːv 'aiən] n. 正离子；阳离子
negative ion['negətiv 'aiən] n. 负离子；阴离子
free radical[ˌfriː 'rædikl] n. 自由基；游离基
onize['aiənaiz] v.（使）电离；离子化
quasi-neutrality['kwaːzi(ː) njuː'træləti] n. 准中性；准中立
laser beam ['leizə(r) biːm] n. 激光束
radiation[ˌreidi'eiʃn] n. 辐射；放射线
dissociate[di'səuʃieit] v. 离解
reactant[ri'æktənt] n. 反应物
etch[etʃ]v. 蚀刻
overlap[ˌəuvə'læp] v. 分重叠；交叠；n. 重叠部分
peel off[piːl ɔf] v. 脱落
destructive[di'strʌktiv] adj. 引起破坏（或毁灭）的；破坏((或毁灭性）的
argon['ɑːɡɔn] n. 氩
nitrogen['naitrədʒən] n. 氮；氮气
nonaqueous ['nəun'eikwiəs] adj. 非水的
slenderize ['slendəraiz] vt. 使成细长状；使苗条
corrosively [kə'rəusivli] adv. 腐蚀地；侵蚀地
keratin['kerətin] n. 角蛋白；角素；角质
erode[i'rəud]v. 侵蚀；腐蚀；风化
helical['helikl] adj. 螺旋的；螺旋形的
folding['fəuldiŋ] adj. 折叠式的；可折叠的
helix['hiːliks] n. 螺旋（形）
configuration[kənˌfiɡə'reiʃn] n. 构象；布局；结构；构造；形状
transformation[ˌtrænsfə'meiʃn] n. 转变；变化；改观
conformation[ˌkɔnfɔː'meiʃn] n. 构造；结构；形态
permanently['pəːmənəntli] adv. 永久地；永远；长期；一直
transform[træns'fɔːm] v. 使改变形态；使改变外观（或性质）；使改观
pleated['pliːtid] adj. 折叠的；打褶的；有褶的
secondary structure['sekəndri 'strʌktʃə(r)] n. 白质的二级结构；次生结构
frictional coefficient['frikʃən(ə)l ˌkəui'fiʃnt] n. 摩擦系数
fracture['fræktʃə(r)] n. 断裂；折断
fracture work['fræktʃə(r) wəːk] n. 断裂功
shrinkable['ʃriŋkəbl] adj. 会收缩的

bulky ['bʌlki] adj. 蓬松的
bulk[bʌlk] v. 膨化；变得巨大；胀大；增大；扩展
contract['kɔntrækt] v.（使）收缩；缩小
relaxation[ˌriːlækˈseiʃn] n. 松弛
specification[ˌspesifiˈkeiʃn] n. 规格
bilateral[ˌbaiˈlætərəl] adj. 双边的；双方的
stabilize['steibəlaiz] v.（使）稳定；稳固
roller['rəulə(r)] n. 罗拉
relax[riˈlæks] v. 松弛
stress [stres] n. 应变
orthocortex [ɔθəuˈkɔːteks] n. 正皮质

Chapter Seven PTT Fiber

polytrimethylene terephthalate (PTT) [ˌpɔliˌtrai'meθili:n ˌterəf'θæleit] n. 聚对苯二甲酸丙二醇酯

generic [dʒi'nerik] adj. 类的；一般的；属的；非商标的

polyethylene terephthalate (PET) [ˌpɔli'eθili:nˌterəf'θæleit] n. 聚对苯二甲酸乙二醇酯

polybutylene terephthalate (PBT) [ˌpɔli'bju:tili:nˌterəf'θæleit] n. 聚对苯二甲酸丁二醇酯

antifouling [ˌætif'fauliŋ] adj. 防污底的；防生物附着的

ethylene ['eθili:n] n. 乙烯

propanediol [prə'pænədiəul] n. 丙二醇

enzymatic [ˌenzai'mætik] adj. 酶催（促）的；酶（性）的

microbial [mai'krəubiəl] adj. 微生物的；由细菌引起的

fermentation [fəːmen'teiʃ(ə)n] n. 发酵

carbohydrate [ˌkaːbəu'haidreit] n. 碳水化合物

monomer ['mɔnəmə] n. 单体；单元结构

ethylene oxide ['eθili:n 'ɔksaid] n. 环氧乙烷

propionaldehyde [ˌprɔpiɔn'ældəˌhaid] n. 丙醛

propylene ['prəupili:n] n. 丙烯

oxidate ['ɔksideit] v. 氧化；n. 氧化物

acraldehyde [ə'krældəˌhaid] n. 丙烯醛

hydroxyl propylaldehyde [hai'drɔksil prəupi'ældəhaid] n. 羟基丙醛

hydration [hai'dreiʃən] n. 水化（合）作用；水化

hydrogenation [ˌhaidrədʒi'neiʃən] n. 氢化（加氢）作用

ferment [fə'ment] v. （使）发酵

melt polycondensation [melt 'pɔliˌkɔnden'seiʃən] n. 熔融缩聚

esterification [eˌsterəfə'keʃən] n. 酯化（作用）

propanediol terephthalate [prə'pænədiəul ˌterəf'θæleit] n. 对苯二甲酸丙二醇酯

transesterification ['trænsəsˌterəfi'keiʃən] n. 酯交换反应

pyrolysis [pai'rɔlisis] n. 高温裂解；热解；高温分解；热裂解

oligomer [ə'ligəmə] n. 低聚物

allyl alcohol ['ælail 'ælkəhɔl] n. 丙烯醇

distillation [disti'leiʃn] n. 精馏；蒸馏；蒸馏法；精华；蒸馏物

esterify [e'sterifai] v. （使）酯化

terephthalic acid (TPA) [ˌterəf'θælik 'æsid] n. 对苯二甲酸

depressurize [di:'preʃəraiz] v. （使）减压；（使）降压

sublimation [ˌsʌbli'meiʃən] n. 升华；升华物

heterogeneous [ˌhetərə'dʒi:niəs] adj. 由很多种类组成的；各种各样的

methoxy [mə'θɔksi] adj. 甲氧基的

dimethyl terephthalate (DMT) [dæ'mi: θail ˌterəf'θæleit] n. 对苯二甲酸二甲酯
dihydroxypropyl terephthalate [hai'drɔksi prəupil ˌterəf'θæleit] n. 对苯二甲酸双羟丙酯
methanol['meθənɒl] n. 甲醇
catalyzer ['kætəlaizə] n. 催化剂
methylene['meθili:n] n. 亚甲基

Chapter Eight　Carbon Fiber

graphite['græfait] n. 石墨
stack[stæk] v. （使）放成整齐的一堆
flake[fleik] n. 小薄片
microcrystal n. 微晶；微晶体
carbonize['kɑ:bənaiz] v. （使）碳化
graphitize ['græfitaiz] v. （使）石墨化
feedstocks ['fi:dstɔk] n. 原料
isotropic [,aisə(u) 'trɔpik] adj. 各向同性的
mesophase ['mesəufeiz] n. 中间相
pitch[pitʃ] n. 沥青
hydrocarbon[,haidrə'kɑ:bən] n. 碳氢化合物
ablate [æb'leit] v. 烧（消，腐）蚀
niche market [ni:ʃ'mɑ:kit] n. 细分市场；利基市场；小众市场
encyclopaedic[en,saikləu'pi:dik] adj. 百科全书般的；知识丰富的
aromatic[,ærə'mætik] adj. 芳香的；有香味的
hexagon['heksəgən] n. 六边形；六角形
turbostratic [tə:bəu'strætik] adj. 乱层的
precursor [pri:'kə:sə] n. 先驱，前辈，前驱物
lubricity [lu:'brisəti] n. 润滑能力
filamentous [filə'mentəs] adj. 细丝状的；如丝的
lubrication[,lu:bri'kein] n. 润滑；润滑作用
carbonization [,kɑ:bənai'zeiʃn] n. 碳化；干馏；碳化物
graphitization [græfitai'zeiʃən] n. 石墨化
size[saiz] v. 上浆
acrylonitrile [ælkrilə(u) 'naitrail] n. 丙烯腈
oxidize['ɔksɪdaiz] v. （使）氧化
homopolymer [hɔmə'pɔlimə] n. 均聚（合）物
copolymer [kəu'pɔlimə] n. 共聚物
thermoplastic [,θə:məu'plæstik] n. 热塑性塑料
trapezoidal [,træpi'zɔdəl] n. 梯形的；成梯形的
decomposition[,di:kɔmpə'ziʃn] n. 分解
disordered[dis'ɔ:dəd] adj. 杂乱的；混乱的；凌乱的
sheath-core[ʃi:θ kɔ:] n. 皮芯；芯鞘；皮芯型
pleat[pli:t] n. 褶
creep[kri:p] n. 蠕变
fatigue[fə'ti:g] n. 疲劳
aluminum[ə'ljuminəm] n. 铝
brittle['britl] adj. 硬但易碎的；脆性的

shear[ʃiə(r)] v./ n. 剪切
impact['impækt] v./ n. 冲击
failure['feiljə(r)] n. 破坏
oxidative ['ɔksideitiv] adj. 氧化的
phosphoric acid[fɔsˌfɔrik 'æsid] n. 磷酸
brittle fracture['britl 'fræktʃə(r)] n. 脆性破裂；脆性断裂
corrosion resistance[kə'rəuʒn ri'zistəns] n. 耐腐蚀性
electromagnetic[iˌlektrəumæg'netik] adj. 电磁的

Chapter Nine　Glass Fiber

quartz[kwɔːts] n. 石英
clay[klei] n. 黏土；陶土
mandrel['mændrəl] n. 心轴（棒）；紧轴；芯模；型芯
platinum['plætinəm] n. 铂；白金
crucible['kruːsibl] n. 坩埚；熔炉；严峻的考验；磨炼
pottery['pɔtəri] n. 陶器（尤指手工制的）；陶土；制陶手艺；制陶技艺
nickel['nikl] n. 镍
chromium['krəumiəm] n. 铬
oxide['ɔksaid] n. 氧化物
alkali-free['ælkəlai friː] n. 无碱
glass fiber reinforced plastics (GFRP) n. 玻璃钢
latex['leiteks] n.（天然）胶乳；人工合成胶乳（用于制作油漆、黏合剂等）
latex cloth['leɪteksklɔθ] n. 胶布
windowscreening ['windəu 'skriːniŋ] n. 窗纱
filter cloth['filtə(r)klɔθ] n. 滤布；滤网
battery isolator['bætri 'aisəleitə] n. 蓄电池的隔离片
magnesium[mæg'niːziəm] n. 镁
silicon['silikən] n. 硅
calcium['kælsiəm] n. 钙
lead [liːd] n. 铅；铅笔芯
silica['silikə] n. 二氧化硅
twistless roving ['twistlis 'rəuviŋ] n. 无捻粗纱
felt [felt] n. 毛毡
batt[bæt] n. 棉絮；棉胎
boric acid[,bɔːrik 'æsid] n. 硼酸
kiln[kiln] n. 窑
rhodium['rəudiəm] n. 铑
sizing agent['saiziŋ 'eidʒənt] n. 浸润剂
coupling agent['kʌpliŋ 'eidʒənt] n. 偶联剂
silicate['silikeit] n. 硅酸盐；硅酸酯
borate ['bɔːreit] n. 硼酸盐，硼酸酯
sodium['səudiəm] n. 钠
potassium[pə'tæsiəm] n. 钾
fluxing agent['flʌksiŋ 'eidʒənt] n. 助熔剂
tetrahedral[,tetrə'hiːdrəl] n. 四面体
trihedral [trai'hiːdrəl] n. 三面体
boron['bɔːrɔn] n. 硼
cation['kætaiən] n. 正离子；阳离子
stress-strain curve[stres strein kəːv] n. 应力—应变曲线

surfactant[səːˈfæktənt] n. 表面活性剂
flexibility [ˌfleksəˈbiləti] n. 柔韧性
bismuth [ˈbizməθ] n. 铋
vanadium[vəˈneidiəm] n. 钒
silicic acid [siˈlisik ˈæsid] 硅酸
coagulate[kəuˈægjuleit] v. 凝固；使凝结
gypsum[ˈdʒipsəm] n. 石膏
anti-skidding[ˈænti ˈskidiŋ] n. 防滑
asphalt[ˈæsfælt] n. 沥青；柏油
cement[siˈment] n. 水泥

Chapter Ten　Aromatic Polyamide Fiber

aramid ['ərəmid] n. 芳纶
polyamide [,pɔli'æmaid] n. 聚酰胺
generic term [dʒə'nerik tə:m] n. 术语；通称；总称
amide ['eimaid] n. 酰胺
aromatic polyamide [ærə'mætik ,pɔli'eimaid] n. 芳香族聚酰胺
poly (m-phthaloyl-m-phenylenediamine) (PMIA) n. 聚间苯二甲酰间苯二胺
poly (p-phenylene terephthamide) (PPTA) n. 聚对苯二甲酰对苯二胺
meta-aramid ['mitə'ərəmid] n. 间位芳纶
para-aramid ['pærə'ərəmid] n. 对位芳纶
liquid crystal ['likwid'kristl] n. 液晶
rheology[ri'ɔlədʒi] n. 流变学；流变性；流变
market introduction['ma:kit ,intrə'dʌkʃn] n. 市场推广
anisotropy [,ænai'sɔtrəpi] n. 各向异性（现象）；非均质性（现象）；有向性
fibrillar ['faibrilə] adj. 纤丝状的，纤维状的
skin-core [skin kɔ:r] n. 皮芯
inception n. [in'sepʃn] n. （机构、组织等的）开端，创始
paramount['pærəmaunt] adj. 至为重要的；首要的；至高无上的
copolyamide [kə,pɔli'æmaid] n. 共聚多酰胺
interpretation[in,tə:prə'teiʃn] n. 理解；解释；说明；演绎
polyamide-hydrazine [,pɔli'æmaid'haidrəzi:n] n. 聚酰胺肼
amine [ə'mi:n] n. 胺；胺类；氨基
carboxylic[ka:bɔk'silik] adj. （含）羧基的
halide['hæ,laid] n. 卤化物；卤素化合物；adj. 卤化物的；卤素的
diamine [dai'eimi:n] n. 二（元）胺
diacid [dai' æsid] n. 二酸（价）；adj. 二酸的；二价酸的
chloride['klɔ:raid] n. 氯化物
p-phenylene diamine [p-'fenəli:n dai'eimi:n] n. 对苯二胺
terephthaloyl chloride [,teref'θælɔil 'klɔ:raid] n. 对苯二甲酰氯
methodology[,meθə'dɔlədʒi] n. 方法，原则
hexamethylphosphoramide (HMPA) n. 六甲基磷酰胺
N-methylpyrrolidone (NMP) n. N- 甲基吡咯烷酮
acetone['æsitəun] n. 丙酮
reaction mixture[ri'ækʃn 'mikstʃə(r)] n. 反应混合物；反应组分；反应（混合）液
blender['blendə(r)] n. 搅拌器
hydrochloric acid [,haidrə,klɔ:rik'æsid] n. 盐酸
stoichiometry [,stɔiki'ɔmitri] n. 化学计算；化学计量学
reactant[ri'æktənt] n. 反应物
reactant mixture [ri'æktənt 'mikstʃə(r)] n. 反应混合物
molar mass['məulə'mæs] n. 摩尔质量；分子量；摩尔量

inherent viscosity[in'herənt vi'skɔsəti] n. 特性黏度
side-reaction[said ri'ækʃn] n. 副反应
N, N-dimethylacetamide n. N, N- 二甲基乙酰胺
mobility[məu'biləti] n. 流动性
gelation [dʒə'leiʃn] n. 胶凝作用；凝胶化（作用）；凝胶的形成（过程）
exothermic[ˌeksəu'θəːmik] adj. 放热的
terephthalic acid [ˌterəf'θælik æsid] n. 对苯二甲酸
calcium chloride['kælsiəm 'klɔːraid] n. 氯化钙
lithium chloride['liθiəm 'klɔːraid] n. 氯化锂
pyridine ['piridiːn] n. 吡啶；氮苯
dicarboxylic acid [daiˌkaːbɔk'silikæsid] n. 二羧酸
diphenyl [daifenil] n. 二苯
trialkyl phosphate[trai'ælkil 'fɔsfeit] n. 三烷基磷酸酯
reagent[ri'eidʒənt] n. 试剂
poly(4-vinylpyridine) n. 聚（4-乙烯基吡啶）
in lieu of[in luː əv] prep. 代替，作为……的替代
substitute['sʌbstitjuːt] n. 代替者；代替物；代用品；v.（以）代替；取代
entropy ['entrəpi] n. 熵
Gibbs energy [gibz'enərdʒi] n. 吉布斯能
dissolution[ˌdisə'luːʃn] n. 溶解
enthalpy [en'θælpi] n. 热函（热力学单位）；焓
of the order ofprep. 大约，与……相似
sulphuric acid[sʌlˌfjuərik 'æsid] n. 硫酸
liquid crystalline phase ['likwid 'kristəlain feiz] n. 液晶相；液晶态
nematic [ni'mætik] adj. 向列型（的）（液晶）
coagulating bath[kəu'ægjuleitiŋ baːθ] n. 凝固浴
lateral['lætərəl] adj. 侧面的；横向的
transverse ['trænzvəːs] adj. 横（向）的；横断的；横切的；n. 横断面
tangential[tæn'dʒenʃl] adj. 切向的
dry-jet wet-spinning n. 干喷湿纺纺丝
lyotropic liquid crystal n. 溶致液晶
circumvent[ˌsəːikəm'vent]v. 设法回避；规避；绕过
hydrazine ['haidrəˌzin] n. 肼；联氨（$NH_2 \cdot NH_2$）
entanglement [in'tæŋglmənt] n. 缠结；纠缠；缠绕物；牵连
therewith [ˌðer'wið] adv. 与此；与之；随即；立即
microvoid['maikrəuvɔid] n. 微孔；微空洞
agitation [ædʒɪ'teɪʃ(ə)n] n. 激动；搅动；煽动；烦乱
superimpose[ˌsuːpəim'pəuz]v. 使重叠；使叠加；使附加于……
truism['truːizəm] n. 不言而喻的道理；自明之理；老生常谈
zigzag['zigzæg] n. 锯齿形线条（或形状）；之字形
hydrogen bonded['haidrədʒən 'bɔndid] adj. 氢键键合的
monoclinic [ˌmɔnə'klinik] adj. 单斜晶系的；单结晶的
orthorhombic [ˌɔːθəu'rɔmbik] adj. 正交（晶）的；斜方（晶系）的

crystallographic [ˌkrɪstələʊˈɡræfɪk] adj.（结）晶的；结晶（学）的
conjugated group [ˈkɔndʒugeitid] adj. 共轭的；成对的
benzamide [ˈbenzəmaid] n. 苯甲酰胺
acetanilide [ˌæsəˈtænlˌaid] n. 乙酰苯胺
terephthalamide [terəfˈθælˈeimaid] n. 对苯二甲酰胺
coplanarity [kəupləˈnærəti] n. 共（平）面性
counteract [ˌkauntərˈækt] v. 抵制；抵消；抵抗
steric hindrance [ˈsterik ˈhindrəns] n. 位阻；空间位阻
analogous [əˈnæləɡəs] adj. 相似的；类似的
dictate [dikˈteit] v. 决定
meridional [məˈridiənl] adj. 子午线的；经线的
periodicity [ˌpiəriəˈdisiti] n. 周期；周期性
processability [prəusesəˈbiliti] n. 可加工性
compressive [kəmˈpresive] adj. 有压缩力的；压缩的
embrittle [imˈbritl] v. 使（变）脆；脆化

Chapter Eleven　UHMWPE Fiber

ultra-high molecular weight polyethylene (UHMWPE) n. 超高分子量聚乙烯
gel-spin [dʒel spin] n. 凝胶纺丝
extended chain [ik'stendid tʃein] n. 伸直链
flex life [fleks laif] n. 弯曲寿命
brand [brænd] n. 品牌
production base[prə'dʌkʃn beis] n. 生产基地
gelate ['dʒeleit] v. 胶凝
lamellar [lə'melə] n. 薄片状的；薄层状的
orthorhombic [,ɔːθəu'rɔmbik] adj. 正交（晶）的；斜方（晶系）的
lamellar crystal [lə'melər 'kristl] n. 片状晶体
monoclinic crystal [,mɔnə'klinik 'kristl] n. 单斜晶体
macrofibril [,mækrəu'faibril] n. 巨原纤
titre ['taitə] n. 纤度
portfolio [pɔːt'fəuliəu] n. 组合
hysteresis loss [,histə'riːsis lɔs] n. 滞后损失；滞后损耗
offshore mooring[,ɔːf'ʃɔː'mɔːriŋ] n. 海中停泊设备；离岸系泊；海上停泊
sheave [ʃiːv] n. 滑轮
wear and tear[weə(r) ənd teə(r)] n. 磨损；损耗
wear and abrasion resistance n. 耐磨性
polyolefin[,pɔli'əuləfin] n. 聚烯烃
cryogenic[,kraiə'dʒenik] adj. 低温的；致冷的
chain slippage[tʃein'slipidʒ] n. 链滑移
thermodynamically ['θəːməudai'næmikəli] adv. 热力学地
coiled conformation[kɔild ,kɔnfɔː'meiʃn] n. 卷曲构象
clinical practice['klinikl 'præktis] n. 临床实践
allergy['ælədʒi] n. 过敏反应
antenna[æn'tenə] n. 天线
fairing['feəriŋ] n. 整流罩
radio transmitter['reidiəu trænz'mitə] n. 无线电发射器；无线电发射机